SPECIAL & GENERAL RELATIVITY

A Beginner's Introduction to Basic & Advanced Concepts

Preetinder Rahil

Limit of Liability/Disclaimer of Warranty

The contents of this book are generic in nature and not directed at anyone specific. The book is not intended for academic purpose. The readers are advised to consult standard textbooks for that purpose. While the author and publisher have made best efforts in preparing this book, no warranties to the accuracy and completeness of the information in this book are made. The internet links provided may not be accurate or appropriate over time, the readers are advised to take due diligence. Neither the publisher nor the author shall be liable for any loss, liability or risk related to the contents of the book, directly or indirectly.

Copyright©2019 by Preetinder Rahil. All rights reserved. No part of this publication may be reproduced, distributed or transmitted in any form or any means, or stored in a database or retrieval system without the prior written permission of the publisher except as allowed under the Copyright Law.

Table of Contents

Preface

Chapter 1 - Review of Mathematics	7
Chapter 2 - Newtonian Gravity	19
Chapter 3 - Special Theory of Relativity	24
Chapter 4 - Curvature of Space-Time	44
Chapter 5 - Space-Time Coordinates	54
Chapter 6 - Relativistic Notation	61
Chapter 7 - Einstein Field Equations	84
Chapter 8 - Black Holes	113
Chapter 9 - Cosmology	147
References	162
Index	164

Preface

Special & General Relativity is Einstein's greatest work and gift to mankind. Relativity is a remarkable theory as it questions the very basis of reality. The nature behaves very differently than what we experience in our daily life. That's why it is even more remarkable how Einstein came up with the theory in the first place. The development of this theory is a shining example of the power of thought. He developed his ideas first from a deep intuitive understanding of nature and then wrote the equations necessary to express those ideas. It should be a guiding principle for today's physicists who develop mathematics first, then ponder over what it all means. It is an exercise in frustration. Relativity is well known in popular culture. Who hasn't heard of $E = mc^2$ equation? Einstein's contribution is much more than this equation, though these days it could be argued that his biggest contribution is the memorable quotes used in social media!

There is no dearth of books on relativity. Einstein himself wrote a popular book on Special and General Relativity. What more do we want? If you can learn directly from the man himself, what am I going to tell you? I am not even a physicist and have no illusions about my intellectual abilities. However there have been plenty of popular and academic books that have been written since then. They all add something, exploring a different perspective that may be useful to a reader. What can I offer?

Most popular books, even the one Einstein wrote do not go into details of the mathematics or the process and intuition behind the equations. They give you a general understanding of the ideas only but not how physics works. Ideas can only tell you so much. It's like showing you a supercar but never taking you for a test drive. Some people may be satisfied with that but not me. There is a satisfaction and happiness to be gained from learning the process and meaning behind equations, after all physics tells us the very meaning of life and nature. But it's not that easy as it sounds. The academic books that offer mathematical rigor and details are too hard for a general reader to understand. The academic books are meant for exams and focus on detailed derivations which may be of much interest to a general audience. There is no easy jump between reading a popular book and going straight to reading a textbook. Try it for yourself, you will know what I mean.

The aim of this book is to bridge the gap between textbooks and popular books. The emphasis of this book is to explain the process and intuition behind relativity equations and explain the implications through the intuitive understanding of them. My hope is after reading this book, when you look at Einstein's field equations, you can say what's going on, why and where do they come from and what do they mean.

How am I going to accomplish this task? Don't you need a strong background in mathematics and physics to learn the subject? The answer is no. There is no doubt that the mathematics is advanced and there is a long list of prerequisites if you want to attend a university level course or read a textbook on relativity. But I think we can get around it. I have no formal physics training and if I can learn the subject despite my pathetic mathematical skills, why can't you? The secret lies in explaining the necessary mathematics in an intuitive manner. I have taken great effort to explain you all the math that you need in a way that is non abstract, meaning easily understandable. The mathematics is best learned in a context and I have extensively used analogies from daily life to give you a feel of the mathematics and equations. Being a non-physicist is an advantage here as I am fully aware of the problems that a general reader experiences as he or she tackles the subject of relativity head-on. I am confident that this approach will be useful for you to learn the subject without spending a lot of time in learning mathematics and then understanding physics, which believe me takes a lot of time and commitment. Who has the time for that? There are no mathematical prerequisites necessary to read this book. Some familiarity with high school mathematics and physics will be useful but I have tried to keep the book self-contained.

I am thankful to all the resources that helped me learn the subject. I have included the list at the end of the book. I welcome any suggestion to improve the book. Please do not hesitate to contact me if you find any errors or inaccuracies in the subject material. I hope you will enjoy reading and learning relativity from the book. A word of caution for students, this book is not designed to replace textbooks or be used for academic purposes. I have tried my best to check the accuracy of the equations but please check with your standard texts before you use any material for academics.

Happy reading!

Chapter 1

Review of Mathematics

There is no escaping mathematics in any branch of physics. But mathematics makes our head heart because our brain is not trained to think abstractly, that is how mathematics is taught. It is much more useful to learn mathematics in a context where we can relate it to our daily life. To learn General Relativity, there is no shortage of mathematical prerequisites. At a bare minimum, you need to have some elementary knowledge of calculus. That's sufficient for this book. If you have never heard of calculus, differentiation, integration and matrices and you want to study General Relativity at the University level stuff which is intended in this book, then I am not sure what has possessed you to study General Relativity. It is like studying Shakespeare without any knowledge of English!

I realize, many of you may have learned calculus in school but that was years or even decades ago and you do not want to go back and learn it again. Who has the time? I won't teach you how to do calculus here but rather give you the intuition behind it so that you can quickly learn the formulae and get going. There are plenty of websites and videos made on elementary calculus.

Why do we need mathematics in physics?

Physics is all about relationships. To objectively, describe a relationship, we need mathematics.
Let's look at an elementary relationship. We know to push anything; we need to apply force. Pushing something means giving it acceleration.

$F \sim a$

Acceleration is proportional to the force applied.

$F = ca$

To get an accurate relationship, we add a constant, that describes a relationship exactly. In the above case, the constant is mass. Heavier objects need more force to push them. How do we know that? Through experience or experiments.

$F = ma$ is of course Newton's second law of motion.

Things are constantly changing; we want to know how things change and more importantly can we predict them.

Where are you now?

In the language of calculus, it is called differentiation.

$\frac{dx}{dt} = now$

d stands for derivative. It means we are interested in your position precisely at a point, not here and there. It is called taking a limit.

$\frac{dx}{dt}$ is called velocity.

How does your velocity change?
It means are you driving on a cruise control or pushing the race pedal or breaking?

$\frac{dv}{dt}$

We want to take the derivative of velocity again. This is called acceleration. There is another way to write it, in terms of position.

$$\frac{d^2x}{dt^2} = a$$

Acceleration is the second derivative of position. That's obvious from the discussion above. We could keep taking the derivatives but in physics, second derivatives are good enough. No need for third or fourth derivatives.

Derivatives can be taken with respect to other variables, not just time. We may want to know not only where you are but what are you doing? Are you jumping or jogging or running?

This means we want to take derivative with respect to position.

$$\frac{df}{dx}$$

f could be any complicated function that determines your enthusiasm to exercise!

There are set formulae to do complicated differentiation. I am just going to list them. Feel free to practice if you feel like but for the purpose of this book, this level of knowledge is sufficient, I hope!

Here are the basic differentiation formulae.

$$\frac{dx^n}{dx} = nx^{n-1}$$

e.g. $\frac{dx^2}{dx} = 2x, \frac{dx}{dx} = 1$

$$\frac{d\,constant}{dx} = 0$$

e.g. $\frac{d\,5}{dx} = 0$

The logic here is simple. If something is constant, how can it change?

The product rule

$$\frac{d(xy)}{dx} = y\frac{dx}{dx} + x\frac{dy}{dx}$$

e.g. $\frac{d(x^2 y)}{dx} = y\frac{dx^2}{dx} + x^2 \frac{dy}{dx} = 2xy + 0 = 2xy$

Note the derivative of $\frac{dy}{dx} = 0$ if y is not a function of x.

The chain rule

$\frac{d(x^2+1)^2}{dx}$, let $u = x^2 + 1$
First differentiate u then the stuff in it.
$\frac{d(u)^2}{dx} = 2u\frac{du}{dx} = 2(x^2+1)\frac{d(x^2+1)}{dx} = 2(x^2+1)2x$

Differentiation of special functions

$$\frac{de^u}{dx} = e^u \frac{du}{dx}$$

$$\frac{de^{x^2}}{dx} = e^{x^2}\frac{dx^2}{dx} = 2xe^{x^2}$$

e is a constant with value of 2.71 and it is the basis of natural logarithm. We will encounter it a lot in physics, so remember how to differentiate it.
It represents exponential growth or decay and its properties are extremely useful.
With exponential growth, as things grows bigger, growth rate increases.

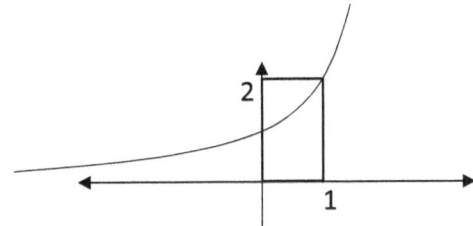

$e^1 = e = 2 \cdot 71, e^5 = 148, e^{10} = 22026$ and so on, it grows fast!

$\frac{d}{dx} sinx = cosx$

$\frac{d}{dx} cosx = -sinx$

Partial Differentiation

There can be variables besides x that we want to calculate their rate of change. The process is pretty simple, just differentiate with respect to each variable. The variable only acts on the function that depends on it.

$\frac{d}{dx} \rightarrow \frac{\partial}{\partial x}$ is symbol of partial differentiation.

$\frac{\partial}{\partial x}(x^2 + 2y^2) = 2x + 0 = 2x$

$\frac{\partial}{\partial y}(x^2 + 2y^2) = 0 + 4y = 4y$

Let's apply partial differentiation to politics.

How would Republicans and Democrats respond to liberal policies like taxing the rich.

$$\frac{\partial}{\partial t} = liberal\ policy$$

$$\frac{\partial}{\partial t}(Republican + Democrat) = 0 + tax\ the\ hell\ out\ of\ them!$$

Integration

Consequences of differentiation is Integration. We are not only interested in where you are now but also where you end up. We want you to show up at work, not sneak around. To do that we integrate all your differentiation steps to get the final answer.

$$v = \frac{dx}{dt}$$

Reversing the process

Displacement(x) = $\int v\ dt$ + K

$\int$ is symbol for integration.

K is a constant and it determines a unique solution. In this context, it may represent your initial position. It is important to know; did you start from home to work or from airport to work. The answer is different in each case.
Let's do some simple examples.

Let $y = x^2$ then $\frac{dy}{dx}$ is

$$\frac{dx^2}{dx} = 2x$$

And the integral is $\int dy = \int 2x dx$

The basic polynomial formula of integration is $\int x^n = \frac{x^{n+1}}{n+1}$

So, $y = 2\frac{x^{1+1}}{1+1} = x^2$

There is a slight problem, the differentiation of $x^2 + 1, x^2 + 2$ etc. is also 2x. So, when we integrate, which equation are we getting?

To avoid this problem, we add constant of integration whose value depends on your choice or what we call initial or boundary conditions.
$y = x^2 + K$

The value of K can be 1,2 or whatever the situation demands. This is why integration is a bit harder than differentiation as solution is not unique.

Finite integral

We may be interested in doing integration between only certain values.
$y = \int_a^b 2x dx$

In this case, we use the fundamental theorem of integration which is
$y = \int_a^b f dx = F(b) - F(a)$
$y = \int_1^2 2x dx = x^2 = (2)^2 - (1)^2 = 3$

It represents area under the curve from a to b. If you are driving a tractor on a farm at certain speed(differentiation) then integration from a to b means how much crop did you sow in the area from a to b.

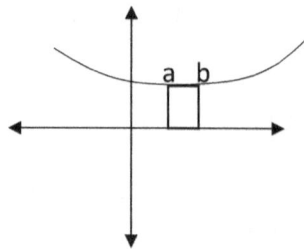

Doing integration is an art. However, the easiest method is-

By substitution

$y = \int \sin(2x + 1)\, dx$

Let $u = 2x + 1$ then

$\frac{du}{dx} = 2$ or $dx = \frac{du}{2}$

$y = \int \sin u \frac{du}{2} = \frac{1}{2}\int \sin u\, du = -\frac{\cos u}{2} = -\frac{\cos(2x+1)}{2} + K$

The constant term is unchanged in an integration.

There are other methods like integration by parts, by fractions and things get pretty complicated. There are integration tables that can be used as well.

Differential equations

The derivatives are mixed up with other functions.

$\frac{dy}{dx} = 5x + 5$

The solution is through integration.

$\int dy = \int (5x + 5)dx$

$y = \frac{5}{2}x^2 + 5x + K$

You can check by differentiating the above equation.

The constant K depends on the initial conditions e.g. if y=2 then x is 4
Will give $2 = \frac{5 \times 16}{2} + 20 + K$

K=-58

$\frac{dy}{dx} - (5x + 5) = 0$

is a first order differential equation.

$\frac{d^2y}{dx^2} - (5x + 5) = 0$

is a second order differential equation.

The second order differential equations are much harder to solve than the first order equations. There are sophisticated techniques to solve the differential equations, but for this book, conceptual understanding is good enough.

Matrices

We need to learn matrix notation. You will see it again and again in General Relativity and Quantum Mechanics. If you know what's happening behind the scene in matrix notation, it simplifies lots of equations. If you do not know what's going on, then you will get lost in General Relativity very quickly.

Let's again start with basic stuff, where are you?

You could be in Vancouver, Montreal or Toronto, because you are rich and rich people own multiple homes!
How should we describe it in matrix notation?

We can form a column vector $\begin{pmatrix} V \\ M \\ T \end{pmatrix}$ and if you are in any one location e.g. Vancouver, that could be described by $\begin{pmatrix} 1 \\ 0 \\ 0 \end{pmatrix}$.

Similarly, if you are in Montreal, that would mean $\begin{pmatrix} 0 \\ 1 \\ 0 \end{pmatrix}$ and for Toronto $\begin{pmatrix} 0 \\ 0 \\ 1 \end{pmatrix}$.

The choice of 1 and 0 to describe position is arbitrary, you could have chosen anything, but 1 and 0 are default choices for obvious reasons.

We can form a row vector as well for position (VMT).

In that case, Vancouver would be represented by $(1\ 0\ 0)$.

The choice to form a row or column vector is arbitrary as well.

We can combine all three choices to form a matrix.

$$\begin{pmatrix} 1 & 0 & 0 \\ 0 & 1 & 0 \\ 0 & 0 & 1 \end{pmatrix}$$

This is a combination of row vector combination.

The diagonal matrix is denoted by $\begin{pmatrix} 1 & & \\ & 1 & \\ & & 1 \end{pmatrix}$

The off-diagonal elements in this matrix are all zero. This simply means that there are no cross terms. This is because, you can only be at one place at a time. If we form a row vector like (1 1 0), then that would have non- zero off diagonal terms, but it is not compatible with the fact that you cannot be in two cities at once.

The matrix can function as an operator, meaning it does something to a test function.

Let's ask a question again, are you in Toronto? $\begin{pmatrix} 0 \\ 0 \\ 1? \end{pmatrix}$

Let's ask the matrix to verify your location.

$$\begin{pmatrix} 1 & 0 & 0 \\ 0 & 1 & 0 \\ 0 & 0 & 1 \end{pmatrix} \begin{pmatrix} 0 \\ 0 \\ 1? \end{pmatrix}$$

This involves matrix multiplication. It means we take each row and multiply by the column vector and then add.

Multiply first row with the column vector then add, $1 \times 0 + 0 \times 0 + 0 \times 1 = 0$

Multiply 2nd row with the column vector then add, $0 \times 0 + 1 \times 0 + 0 \times 1 = 0$

Multiply 3rd row with the column vector then add, $0 \times 0 + 0 \times 0 + 1 \times 1 = 1$

$$\begin{pmatrix} 1 & 0 & 0 \\ 0 & 1 & 0 \\ 0 & 0 & 1 \end{pmatrix} \begin{pmatrix} 0 \\ 0 \\ 1? \end{pmatrix} = \begin{pmatrix} 0 \\ 0 \\ 1 \end{pmatrix} \text{ Yes, you are in Toronto!}$$

Addition and subtraction of matrices is straight forward. Just add or subtract each entry.

$$\begin{pmatrix} 0 \\ 0 \\ 1 \end{pmatrix} + \begin{pmatrix} 0 \\ 0 \\ 1 \end{pmatrix} = \begin{pmatrix} 0 \\ 0 \\ 2 \end{pmatrix}$$

$$\begin{pmatrix} 0 \\ 0 \\ 1 \end{pmatrix} - \begin{pmatrix} 0 \\ 0 \\ 1 \end{pmatrix} = \begin{pmatrix} 0 \\ 0 \\ 0 \end{pmatrix}$$

Do you want to see non zero diagonal matrix?

Let's ask a question, where do you find traffic jams?

$(VMT) = (1\ 1\ 1)$

The matrix will become

$\begin{pmatrix} 1 & 1 & 1 \\ 1 & 1 & 1 \\ 1 & 1 & 1 \end{pmatrix}$ which means traffic jams are everywhere!

Sometimes you see trace of a matrix mentioned. It is a simple concept. Simply add diagonal elements of the matrix. It is useful if entries are probabilities of events. In our example of finding you in three cities, assuming you are to be found equally likely in all three cities, matrix will become

$$\begin{pmatrix} \frac{1}{3} & 0 & 0 \\ 0 & \frac{1}{3} & 0 \\ 0 & 0 & \frac{1}{3} \end{pmatrix}$$

Trace of the matrix is $\frac{1}{3} + \frac{1}{3} + \frac{1}{3} = 1$

It means there is 100% probability of you being found in three cities.

This is matrix introduction for you, not bad eh!

We are now ready to move on to physics.

Chapter 2

Newtonian Gravity

The theory of gravity developed by Newton is simple, elegant and powerful. It is simple enough to be taught in schools and powerful enough to be used for space exploration. The theory is very accurate in describing gravity when the gravitational field is weak, and things are moving slowly as compared to the speed of light. In physics jargon it is called a non-relativistic theory of gravity. It works on a simple assumption that everything attracts everything. A physical mass exerts a force on another mass which is given by **Newtonian law of gravitation**

$$F = \frac{G m_1 m_2}{r^2}$$

Where m_1 and m_2 are the masses of physical bodies, r is the distance between them and G is the gravitational constant.

The value of G is $6.67 \times 10^{-11} \ m^3 kg^{-1} s^{-2}$.

The value of G is determined by experiments, which is true for any constant that we see in physics. But the gravitational constant along with Planck's constant and speed of light are considered fundamental to our existence in the universe.

The direction of force is towards each mass. It means the two bodies attract each other with equal and opposite force.

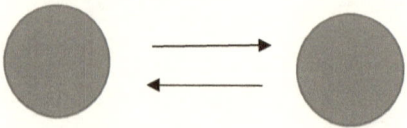

This is basically Newton's third law of motion.

We also know from Newton's second law of motion that force is equal to mass times acceleration.

$F = ma$

Equating both the equations leads to

$m_1 a = \frac{G m_1 m_2}{r^2}$

If we consider one mass to be earth (m_2) and the other mass (m_1) to be an object falling freely under gravity, then the acceleration due to gravity is

$g = \frac{G M_{earth}}{r^2}$

Notice that mass of the falling object cancels out. This has profound implications. First of all, it assumes inertial mass is equivalent to gravitational mass. The inertial mass is what goes in $F = ma$ equation. It creates resistance to its state of rest or motion. The gravitational mass is what goes in Newton's law of gravitation. It means whether a body is heavy or light, it falls with the same gravitational acceleration as long as the air resistance is not high. This was what Galileo did when he threw objects of different mass from the leaning tower of Pisa. It is a different matter if it actually happened or it was just a thought experiment. The other implication is that it does not matter what a mass is made of. A cotton ball and a steel ball of the same mass will fall with the same acceleration. This is called the weak equivalence principle. Einstein took this principle to the next level and it founded the basis for the General Theory of Relativity. It is a different matter that his thought process was entirely different. We will explore this principle in detail later.

There is a problem with Newton's law of gravitation. It assumes that gravity acts instantly between bodies. A rock on earth is exerting gravitational force on the most distant stars in the galaxy instantly! This does not seem right. Once it was found that speed of light is the universal speed limit and things don't just happen instantly across the universe, it prompted Einstein to build his Special and General Theory of Relativity. The Newton's gravitational theory does not answer how exactly gravity works. What causes the force of gravity between masses? It only gives a formula to calculate its value, which by no means is a small feat.

The concept of gravitational field is a way to explain how gravity acts at a distance but still leaves the underlying mechanism unexplained. The concept of field is simple. It means area of influence on others. When an exotic car is cruising on a street, it turns heads but there is an area in which this happens. A person on another street who cannot see the exotic will not turn his head! He is outside the attractive field of the car.

The gravitational field extends forever. The technical definition is the force experienced by a test mass.

$$g = \frac{F}{m}$$

In case of earth, $g = -\frac{GM_e}{r^2}$
The minus sign simply means its attractive, as force is towards the earth, not away from it. The force acts in the direction opposite to the displacement away from the earth.

There is another way to describe the gravitational field, through Gauss's law. The Gauss's law in case of an electric field is $\phi_E = \frac{Q}{e}$.

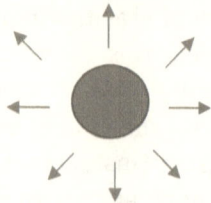

The flux of the electric field which determines how strong the field is, depends on the strength of the charge inside a closed surface or volume. That's common sense, a river flowing from a glacier has more flux than a tap water flowing from a water tank.

The Gauss's law for gravity states that the mass inside a closed surface determines the strength or flux of gravitational field.

$$\nabla . g = -4\pi G\rho$$

∇ is called the divergence. It means how the lines of gravitational field are spreading out. It is given by calculating $\frac{\partial}{\partial x}, \frac{\partial}{\partial y}, \frac{\partial}{\partial z}$ .So, in each direction, small changes to the direction of field lines are tracked. In case of gravity, the lines are converging, that's why there is a minus sign.

ρ is the mass density. We can use this theorem to derive the gravitational field that we calculated earlier. In case of a spherical body like earth where surface area is given by $4\pi r^2$, the law will give

$$-4\pi GM = 4\pi r^2 g \text{ or}$$

$$g = -\frac{GM_e}{r^2}$$

Let's move on to the gravitational potential energy. Traditionally it is defined in terms of the work done in moving a test mass from one point to another (a to b).

Potential energy change = $\int_a^b -F \cdot dr$

Now, we already know the gravitational force formula which is $\frac{Gm_1m_2}{r^2}$.

The integral will become $\int_a^b -\frac{Gm_1m_2}{r^2} \cdot dr$

If we take the initial point to be infinity which is at distance $r = \infty$, then the formula for potential energy resolves to $-\frac{Gm_1m_2}{r}$.

The concept of field can be extended to potential energy as well. The potential energy per unit mass (Φ) defines the field. In case of the earth, it will be $-\frac{GM_e}{r}$.

The potential can be written in another way, $g = -\nabla\Phi$. Putting this value in the Gauss's law formula leads to

$\nabla(-\nabla\Phi) = -4\pi G\rho$ or

$\nabla^2\Phi = 4\pi G\rho$

It is an important equation, which will be used later in developing General Relativity. ∇^2 is called the Laplacian operator given by $\frac{\partial}{\partial x^2}, \frac{\partial}{\partial y^2}, \frac{\partial}{\partial z^2}$.

That's enough Newtonian gravity we need to know before moving on to the next level.

Chapter 3

Special Theory of Relativity

Special theory of relativity was developed first. Einstein was working as a clerk in the Swiss town of Bern in 1905.He published a series of papers in 1905 that changed the world of physics forever. 1905 is called the miracle year or Annus mirabilis. $E=mc^2$ we all know, is part of special relativity. Einstein's reputation was cemented by profound implications of this theory. Special relativity is based on very simple principles, but consequences are far reaching.

The core principles of special relativity are
1. Laws of physics are same for all inertial observers.
2. Speed of light is same for all inertial observers.

That's it. With these two principles, whole theory can be built. Isn't it brilliant?
The first principle is easy to understand. The inertial observers mean the observers can have different speed with respect to each other. So, one observer can be standing on the road, another one is going in a car with constant speed. Both are in inertial frame. Whether you are doing an experiment in a lab at rest or in a lab moving at constant speed, the laws of physics do not change. This makes intuitive sense. Non-inertial frame means acceleration or deceleration is involved. This messes up laws of physics. If the lab suddenly accelerates, one will feel a pseudo force backwards and the equipment can fly around! So non-inertial observers do not experience same forces.
Inertial frame of reference= rest or moving at constant speed→ Laws of physics same.
Non- inertial frame of reference = acceleration or deceleration → Laws of physics not same.

The second principle is a curious one. It was not apparent that speed of light should be same for all inertial observers. In fact, it does not make intuitive sense.
The observer in a car moving at 100 miles an hour, shoots a beam of light and measures its speed.
The moving observer will measure speed of light= c.
The observer at rest is also measuring the speed of light at the same time.
So, rest observer should measure speed of light= c + 100 miles an hour
This is because the light was already moving at 100 miles an hour with respect to the rest observer, when light beam started.
The experiments have shown that both rest and moving observers measure the same speed of light = c.

Michelson -Morley experiment in 1887 provided the convincing evidence that speed of light is same for all inertial observers. It was designed to check for the presence of ether. It was thought that light needed a medium called ether to travel. So, as earth rotates, the ether moves with respect to the earth. The speed of light should be more when traveling along ether and less when traveling against the direction of ether. If you move in the direction of blowing wind, you have wind at your back!

The experiment split the beam of light in different directions and combined it later by mirrors. It did not find any difference in the speed of light in a given direction.
If speed of light is same for all observers, then what gives?

$$\text{Speed} = \frac{distance}{time}$$

So, our notion of distance and time has to change to keep the speed of light same. Some may argue that the constancy of speed of light is a consequence of the principle that laws of physics are same for all inertial observers. If speed of light is not constant for inertial observers, then Maxwell's equations for electricity and magnetism that incorporate speed of light will be different for inertial observers, which is not allowed.

Consequences of special relativity

1. *Violation of simultaneity*

The events that are simultaneous in one frame, may not be simultaneous in another frame.
The classic example is a person traveling on a train. He shines a beam of light that travels towards front and back of the train compartment at the same time. The person on the train will see both beams of light striking front and back of the train simultaneously.

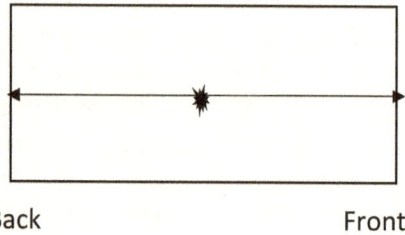

Back   Front

The observer on the train platform who is at rest, will see things differently. He will see beam of light hitting back of the train first, followed by front of the train. This is because back of the train is coming towards beam of light, so their relative speed is high. The front of the train is moving away from beam of light, so it gets hit later. So, the events that are simultaneous for the observer on the train are not simultaneous for the observer on the ground
The events that are connected by light signal are casually related. When an event happens, say an explosion happens at time zero, it spreads outside with a maximum limit of speed of light.

Or it spreads like a cone from one time to another.

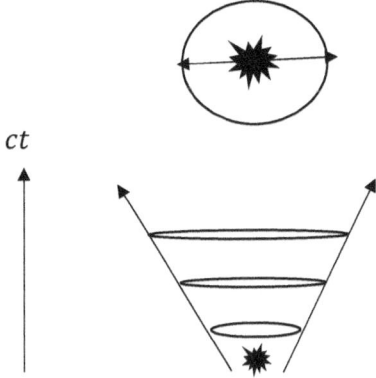

ct

The time flows upwards and is usually combined with speed of light. This is to make units consistent as ct has units of distance. The cone formed by the flow of events is called the light cone.

The graph of the cone looks like this

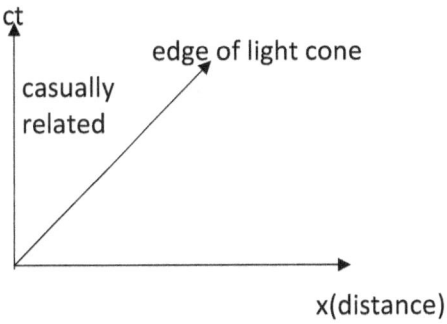

The edge of the cone refers to the angle of 45 degree. On this line, the events move at the speed of light. Below it, the distance travelled increases, so speed can exceed the speed of light. The events that take place below this line or outside the light cone are not casually related. The events that take place above the light cone line travel at less than the speed of light and are casually related.

The cause and effect are not violated as all observers agree on the order of events inside the light cone

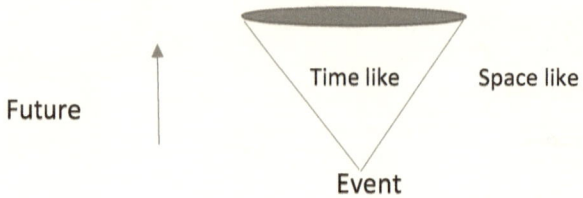

Future

The cone in the upwards direction depicts future events. We can draw a cone facing downwards from the event, which will depict the past with respect to the event.

The events inside the light cone are casually related and called time-like. This is because there is a frame of reference in which events happen at the same place but at different times. The bullet firing from a gun takes place at the same place at different times so these events are casually related and time like.

The events outside the light cone are not causally related and called space like. This is because there is a frame of reference where events happen at different places at the same time. This makes some sense, can someone sitting on a beach see a bullet being fired at one ship and someone die at another ship at the same time? These two events cannot be causally related as bullet firing and reaching a person on another ship has to happen instantly, breaking the speed of light limit which is not allowed.

The moving frames may be tilted but all frames and observers agree on time like events inside the cone.

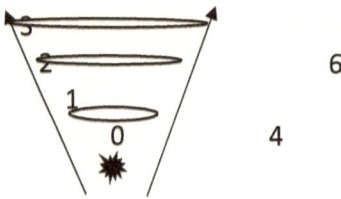

The events 0,1,2 and 3 are causally related as they are inside the light cone. They could represent

Event 0 = bullet fired

Event 1= hit a person

Event 2= CPR is performed

Event 3 = person dies

Event 4 = NASA probe on the edge of galaxy stops working

Event 6= star explodes

All observers will agree on the events inside the light cone. The person cannot die of bullet injuries and bullet hits him afterwards!

The line that connects events is called the world line. The space time diagrams shown are called **Minkowski** diagrams. They represent flat surface as we have not considered curved space time that comes into picture when doing General Theory of Relativity.

The distance between two events(s^2) is frame independent.

$S^2 = c^2 (t_1 - t_2)^2 - (x_1 - x_2)^2$

$S^2 > 0$ is time like, inside the light cone

$S^2 = 0$ is light like, on the light cone line

$S^2 < 0$ is space like, as $x_1-x_2 > t_1-t_2$ and speed will exceed speed of light.

Where does this formula come from?

We have all learned about Pythagoras theorem that tells us about distance between two points.

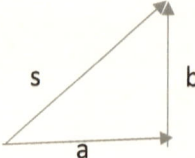

$$S^2 = a^2 + b^2$$

Why is space time distance formula different?

It is centered around the speed of light being the connection between cause and effect. If time is clicking more and distance between events is less than they are casually related. This is the whole premise of the world line; things are causally related so that series of events make sense.

Going back to the example of the train, the light reaching the end of the compartment could trigger a reaction. If the light reaching the end of the compartment causes the person at each end to fire a bullet, then the observer on the train will see that the persons on the front and back end of the compartment fired bullets at each other simultaneously. The observer on the ground will see that the person on the back end of the compartment fired first. This is because the events at the end of the train are space like, outside the light cone so observers can disagree on the sequence of events. But the event where torch started and caused the beam of light to travel in each direction is time like and all observers will agree on the fact that torch button was pushed first, and then light beam started, not the other way around. Note that the action of pressing a button and light coming out of the torch happened at the same place but at different times, so it is a time like event. The light reaching ends of the train happens at different places at the same time in one frame, so it is space like event.

Do not worry about guns and bullets in the examples, physicists are peace loving individuals.

2. Time Dilation

It is hard to get your head around that time is relative. We all have a notion of time and our minds have been wired to think that 1 second on earth is 1 second everywhere in the Universe. The notion of absolute time is the basis of Newtonian mechanics.

But in relativity every place has a special clock and different observers can disagree on time.

Let's go to the example of the moving train. Most of the relativity takes place on trains! In physics there is a term for these kinds of experiments. They are called Gedanken or thought experiments.

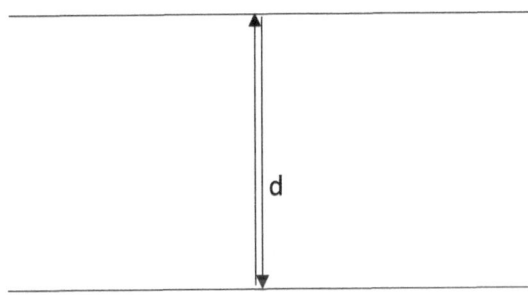

A light is shined from the bottom of the train carriage. The beam of light goes up, hits the ceiling and comes back at the same place. The observer on the train will see the beam travel a distance of 2d and take time t to travel that distance.

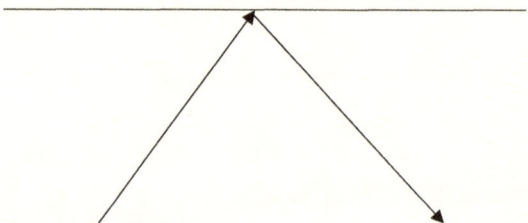

The land-based observer will see a different trajectory of light. It will not be back and forth in a straight line. It will hit the ceiling at an angle and come back at an angle as by the time the light reaches the ceiling, the train has moved by velocity v.

We can draw a triangle and use Pythagoras theorem to calculate the time for land-based observer.

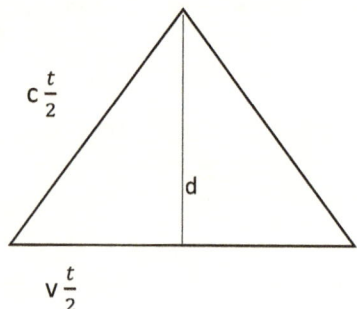

The light covers distance ct and train covers distance vt.
Using Pythagoras theorem, we have
$$\left(\frac{ct}{2}\right)^2 = \left(\frac{vt}{2}\right)^2 + d^2$$
Or
$$T_{land} = \frac{2d}{c\sqrt{1-\frac{v^2}{c^2}}}$$
$$T_{train} = \frac{2d}{c}$$
We define $\gamma = \frac{1}{\sqrt{1-\frac{v^2}{c^2}}}$

$T_{land} = \gamma \, T_{train}$

γ is always more than 1 as v cannot exceed c in the denominator term.

This means time as calculated by the land-based observer, will be longer than the time observed by the observer on the train. The land observer will say that the things are happening slowly on the moving train. This is time dilation. The land-based observer will say "moving clocks run slow".

The peculiar thing is that the person who is in the frame does not notice time is fast or slow, time is running normal, like clockwork!

The train person has no idea that land-based observer is finding his clock running slowly as when he checks his watch, it is keeping perfect time.

If the train was moving at the speed of light, then the denominator will be zero and time as observed by the land observer will be infinite. The train will appear frozen in time. It is a different matter that massive particles cannot reach speed of light as it requires infinite energy, so only massless particles like photons travel with the speed of light.

Who is moving with respect to whom?

If train is moving with respect to land observer, same can be said by the train observer that the land person is moving, and he is stationary. Now on earth this comparison is not quite valid as train has to accelerate from rest.
But take the case of an asteroid and earth, observers on both can say that one is stationary, and it is the other who is moving.
Earth observer will say asteroid clock is slow and asteroid observer can say the earth clock is slow. Who is right?

Both are right, time is relative.

It is like an argument; everyone thinks they are right!

Relativity of time can lead to interesting paradoxes.

The twin paradox is the classic one.

If twins get separated, one travels in a space ship to galaxy far away. The other twin stays on earth. The spaceship is capable of travel near the speed of light. The twin comes back after spaceship journey. It takes 10 years for him, but when he returns to earth, he finds that the earth twin has aged 40 yrs and is an old man.
This is because his clock was slow and earth clock was fast.

But speed is relative, it could be argued that it is the earth that moved away at fast speed from the space ship and so the earth clock should be slow and earth twin should be younger, not older than him.

This is the paradox. The answer is not complicated. The space twin is not in inertial frame all the time. The space ship has to accelerate to reach high speed and then turn back to reach earth. Speed and time are only relative in inertial frames. Non-inertial frames are special frames where other forces come into picture. So, it is the space twin who is younger and the poor earth twin who gets older.

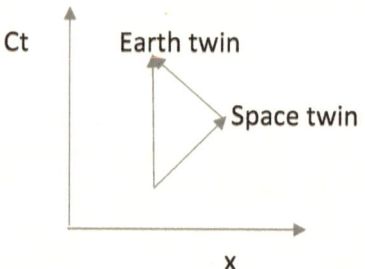

The world lines are an easy way to resolve the twin paradox. The earth twin's world line is straight up whereas space twin's world line is tilted. It may appear that the earth twin's world line is smaller than the space twin's line but it's not true. The earth twin's line is straight up where the time coordinate is maximized. The space twin's world line tilts towards space coordinate and away from time coordinate. So, the space twin has less time coordinate in his world line.

To avoid confusion, the concept of proper time is invented. Proper time is the time observed when at rest with respect to the event. In the train example, proper time is the time observed by the train person as torch is lit on the train.

$T_{land} = \gamma \, T_{proper}$

Some books use τ (tau) to denote proper time.

Time dilation is not a fancy, out of reach concept, it has practical implications.

GPS system is based on corrections to time change based on gravity. Muon has a short life span or proper time. Since muon travels at relativistic speeds,

$T_{earth} = \gamma \, T_{muon}$

The muon life span is increased due to time dilation on the earth. So, muons last long enough to reach the earth lab from the atmosphere.

Sitting in a mathematics class feels like an eternity but playing videogames all day takes no time at all!

3. Length contraction

How do we calculate length?

If we want to calculate distance between earth and a star, we can send a space ship from earth to the star. The space ship will calculate the distance it has traveled, which will be the distance between earth and the star.

Length as calculated by spaceship = v (speed of the space ship) × time taken by the spaceship as per spaceship clock
$L_s = v \times t_s$

The earth-based space station can also calculate the distance

Length as calculated by earth= v × time taken by the spaceship as per earth clock

$L_e = v \times t_e$

Now, $t_e = \gamma t_s$

$L_s = v \times \dfrac{t_e}{\gamma}$

Or $L_s = \dfrac{L_e}{\gamma}$

The length measured by moving observer is contracted as compared to length measured by the observer at rest to the measured quantity.

L_e is the proper length as it is measured in the rest frame of the measured quantity. Both proper time and length are calculated in the rest frame of the measured object. The proper measurements are done at rest with patience, not in a hurry running around.

Length paradox

The length contraction by moving objects can lead to interesting paradoxes.

Imagine there is a fighter pilot with the state-of-the-art aircraft capable of speed close to the speed of light. The pilot sees a small opening in the bunker on the ground in the enemy territory. He has a bomb which is 1 m in length. He calculates the length of the bunker opening but finds its only 0.7m. He is not happy. The bomb will not go in the bunker. The commanding officer on the ground tells him that opening is 1.5 m and he can get the bomb through. The pilot is ordered to release the bomb and he follows through with the order.

What will happen?

Will the bomb go through the bunker or not?

Who is correct?

Both are correct. The length contraction as measured by the pilot has caused the discrepancy in the results.

The command center will see the bomb go through the bunker.

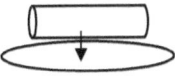

The pilot will also see the bomb go through the bunker, but his calculation is still correct. He will see the bomb slide though the opening.

This again shows the violation of simultaneity. The land observer sees both ends enter the bunker at the same time, but the pilot sees one end going first.

Coordinate Transformation

How quantities change with change of coordinates or frames of reference is the name of the game.

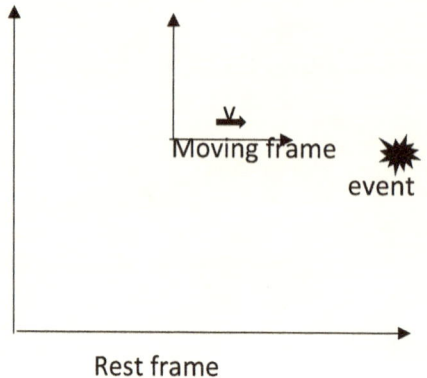

The moving frame has a velocity v. The event could be at rest or moving with velocity v.

The transformation in Newtonian mechanics is pretty straight forward.

If distance of the event from the origin is x in the rest frame, then in the moving frame, distance x' will be
$x' = x - vt$
The other coordinates remain unchanged as movement is only in the x direction
$y' = y$
$z' = z$
Time is universal in Newtonian mechanics
$t' = t$
These are called Galilean transformation equations.

Lorentz Transformations

Transformation equations need to be modified for special relativity due to time dilation, length contraction and speed of light being constant for all observers. The equations are called Lorentz transformations.

$$x' = \gamma(x - vt)$$

$$y' = y$$

$$z' = z$$

$$t' = \gamma\left(t - \frac{v}{c^2}x\right)$$

$$u' = \frac{u_x - v}{1 - \frac{u_x v}{c^2}}$$

Here v is the speed of the moving frame and u is the speed of the event.

If the event is moving with speed of light, say beam of light was started at rest, then the moving frame will observe the same event with speed

$$u' = \frac{c-v}{1-\frac{cv}{c^2}} \text{ or } \frac{c-v}{1-\frac{v}{c}} \text{ or } \frac{c\left(1-\frac{v}{c}\right)}{\left(1-\frac{v}{c}\right)} = c$$

This agrees with the basic tenet of special relativity that all observers agree on the speed of light.

Lorentz invariance is the quantity that is unchanged by Lorentz transformation. It is worth in gold in special relativity.

The space time distance between two events ($s^2 = t^2 - x^2 - y^2 - z^2$), speed of light and electric charge are some examples of Lorentz invariance.

It is like saying that observers will disagree on who is moving with what speed, but they will agree on the color of their clothes and the smell of their scent.

Relativistic momentum

It is not too hard to figure out.

Classical momentum is $m \frac{dx}{dt}$. But which time should we use? Answer is the proper time.

$\frac{dx}{d\tau}, \frac{dy}{d\tau}$ and $\frac{dz}{d\tau}$

But time in a chosen frame is $t = \gamma \tau$ or $d\tau = \frac{dt}{\gamma}$

So, relativistic momentum, $p = m \frac{dx}{d\tau} = m \gamma \frac{dx}{dt}$

Or $p = \gamma\, mu$

Where u is the velocity.

Relativistic Energy

To get to the energy, we first derive the work done to move an object.

$W = \int Force \cdot dx = \int \frac{dp(momentum)}{dt} dx$

$dx = u\,dt$, where u is the velocity. Substituting it in the equation

$\int \frac{dp}{dt} u\,dt = \int_0^v u\,dp$

The integral can be solved by doing integration by parts

$up - \int p\,du$

Substituting the value of relativistic momentum

$$\frac{mu^2}{\sqrt{1-\frac{u^2}{c^2}}} - \int \frac{mu}{\sqrt{1-\frac{u^2}{c^2}}} du$$

By inspection or guess work, the integral can be solved to

$$\frac{mu^2}{\sqrt{1-\frac{u^2}{c^2}}} + \left| mc^2\sqrt{1-\frac{u^2}{c^2}} \right|_0^v$$

We can also put integration limits from 0 to v

$$\frac{mv^2}{\sqrt{1-\frac{v^2}{c^2}}} + mc^2\sqrt{1-\frac{v^2}{c^2}} - mc^2$$

The term $mc^2\sqrt{1-\frac{v^2}{c^2}}$ can be multiplied by $\frac{\sqrt{1-\frac{v^2}{c^2}}}{\sqrt{1-\frac{v^2}{c^2}}}$

We will get $\frac{mc^2 - mv^2}{\sqrt{1-\frac{v^2}{c^2}}}$

So, we have

$$\frac{mv^2}{\sqrt{1-\frac{v^2}{c^2}}} + \frac{mc^2}{\sqrt{1-\frac{v^2}{c^2}}} - \frac{mv^2}{\sqrt{1-\frac{v^2}{c^2}}} - mc^2$$

Finally, we get the expression

W= $\gamma mc^2 - mc^2$

This work done will get converted into kinetic energy of an object as we are taking an object at rest and giving it momentum p.

KE = $\gamma mc^2 - mc^2$

This does not look like ½ mv², the equation that we come to know from high school. Be patient, first expand γ term in Taylor series expansion

$$\gamma = \frac{1}{\sqrt{1-\frac{v^2}{c^2}}} = 1 + \frac{1}{2}\frac{v^2}{c^2} + \text{ignore higher terms}$$

Taylor series is like any other power series where higher terms are added to increase the accuracy. The exact formula can be found in textbooks.

Basically, taking first few terms of Taylor series is like me drawing Mona Lisa as :)

We need lot of higher terms to make it look like real Mona Lisa but in physics, higher terms become less and less important. Basic approximation will do.

Then Kinetic Energy = $mc^2 (1 + \frac{1}{2}\frac{v^2}{c^2}) - mc^2$

Mc^2 terms cancel out and we are left with $\frac{1}{2}mv^2$.

How to interpret the right-hand terms?

γmc^2 is Total Energy= E

mc^2 is the rest energy.

E = Kinetic Energy + rest energy(mc^2)

If momentum is zero, then

E= mc^2

The most famous physics equation ever!
Note that if you make speed of light =1, the natural units
Energy and mass are equivalent. They can be converted into one another.

Energy Momentum Relation

Energy = γmc^2

Squaring $E^2 = \gamma^2 m^2 c^4$

Momentum, $P = \gamma mv$

Multiplying both sides by c and squaring

$c^2 p^2 = c^2 \gamma^2 m^2 v^2$

Subtracting it from E^2 term we get

$E^2 - c^2 p^2 = \gamma^2 m^2 c^4 - \gamma^2 m^2 v^2 c^2 = \gamma^2 m^2 c^4 \left(1 - \frac{v^2}{c^2}\right) = \gamma^2 m^2 c^4 \times \frac{1}{\gamma^2} = m^2 c^4$

Or $E^2 = c^2 p^2 + m^2 c^4$

This is called *energy-momentum relation*.

For massless particles

$E = pc$

Note that $E^2 - p^2 = m^2$ is an invariant term. So, all observers agree on the invariant mass.

The invariant mass is independent of the motion of the object. It is the total mass or energy of the object in the rest frame.
In our train analogy, observers will not disagree on the number of people in the train if they rely on the camera that was installed in the train in its rest frame.

The special relativity needs relativistic notation to make things relatively easy to solve. We will earn about relativistic notation in subsequent chapters.

Chapter 4

Curvature of Space - Time

What do we mean by curvature? Most people have some sense of curvature. If an object has bends or curves, it has curvature. But that's not good enough for physics. The curvature we normally describe is called the extrinsic curvature. By extrinsic, I mean the curvature is described from the outside. We look at a book or a painting from the outside at a distance. The technical way of saying extrinsic curvature is that we are analyzing how an object is embedded in a higher dimensional space. In relativity, extrinsic curvature is not that important. It is the **intrinsic curvature** that is paramount. The reason is that the universe or the spacetime we live in, we are inside it. There is no way to analyze it from outside. For that we have to go to outside the universe to see how curvy it looks. That's not possible!

Which one of the above is curved?

The answer is none of the above. The intrinsic curvature for a line is not defined. An object living on a one-dimensional line has no concept of curvature. What we perceive as a curved line is by embedding in the two dimensions or in other words, looking from outside. We are three dimensional beings; we cannot really imagine living in a one dimensional or two-dimensional world. Only mathematics can give us some idea how to analyze the intrinsic curvature of spaces.

Moving on to two dimensional spaces. Take a piece of paper, fold it into an aero plane or a ship. Does that make it curved? Obviously not. It is still flat, folding it into a ship is just embedding it in higher(three) dimensional space. We can easily unfold the ship and get our flat paper back. The distance between two points do not change by folding it. A quick test for flat curvature is, if something can be flattened on a plane or your desk, without distorting it, then it does not have intrinsic curvature. We can easily unfold and flatten a piece of paper, regardless of its folding pattern. This means the piece of paper has no intrinsic curvature.

Let's go to our three-dimensional world. Take a tennis ball. Does it look curved to you? Well, let's do our quick flatness test. Try to flatten it on a flat desk. To do that, you have to cut the ball into two and try to flatten it. What will you find? The ball will get distorted and stretched in your attempt to flatten the ball's surface onto the desk. This means it has real intrinsic curvature. We will study the use of coordinates later but in terms of coordinates, the flatness test means that if Cartesian coordinates can be used to describe an object, then it is not curved. The flat desk can be described by x-y plane, which is a Cartesian or a flat coordinate system. A ball with real intrinsic curvature cannot be described by flat coordinates.

A crucial test for flatness is how directions change as we move along a curved surface. Intuitively, the directions should change as we go along a curved path. A vector that represents direction of motion is transported along a closed loop. We have to keep the direction and magnitude same as vector goes along the curve. To do that, make sure vector stays parallel to the previous position. This is called **parallel transport**. This is the gold standard for checking intrinsic curvature.

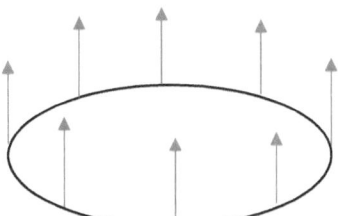

As you can see, parallel transporting a vector in a loop on a flat surface, does not change the direction of the vector.

In parallel transport change of direction is not allowed.

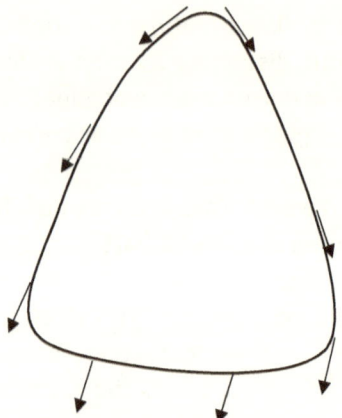

As you finish going around the loop, you will find that the last vector will be at an angle with respect to the starting vector.

θ is the geometric phase angle. This represents curvature. To see the true effect of parallel transport, you have to do it over a three-dimensional sphere. On a book, I cannot show you the real curvature. The embedding can work in the opposite direction too, meaning three-dimensional world can be mapped on to a two-dimensional surface. This is nothing new, maps and paintings do it all the time. But embedding does not give you the intrinsic curvature, you have to rely to mathematical concepts like parallel transport. Let's look at curves below

Which one has more curvature?

The smaller curve is more curved. To see that, pick a point on the curve, draw a circle about that point and draw tangent on opposite sides, intersecting as close as possible to that point.

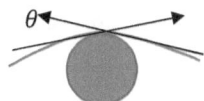

θ is more in case of the smaller curve. It is related to the radius of curvature by $\theta = \frac{1}{R}$. The curvature k is usually defined as $k = \frac{1}{R}$. So, bigger the circle, smaller its curvature. That's why earth's curvature is much smaller than the curvature of a ball. It is not surprising people believed in flat earth for so long. We have analyzed these curves on a flat plane as in one dimension, curvature is not defined.

The curvature can be positive or negative. We do have some intuition about positively curved or negatively curved surface. We can naively say that convex surface is positively curved and concave surface is negatively curved.

But again, we are embedding the curves in two-dimensional plane and convexity is thus arbitrary.

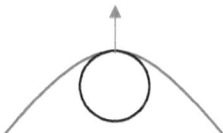 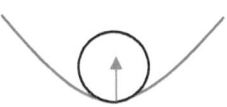

Let's take example of a curved surface where the above curves are part of it. We can draw a circle around a chosen point. We can also draw a vector which is perpendicular to that point. If the circle is on opposite side of the vector then curvature is positive, if on the same side, its negative. We have to play this game by rotating around the point. This can be done if normal vector is part of the plane that can be rotated around the chosen point. This way various values of curvature k can be calculated. The maximum and minimum values of k define the Gaussian curvature.

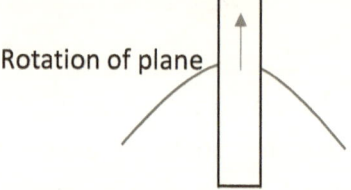

$$K = k_1 k_2$$

k_1 and k_2 are the maximum and minimum values. They are called principal curvatures.

If K is positive, that defines positive curvature. A sphere has positive curvature.
If K is negative, that defines negative curvature. A saddle has negative curvature.
If K is zero, that defines zero curvature. A cylinder has zero curvature.
Gaussian curvature is a measure of intrinsic curvature, that's why it's very useful.

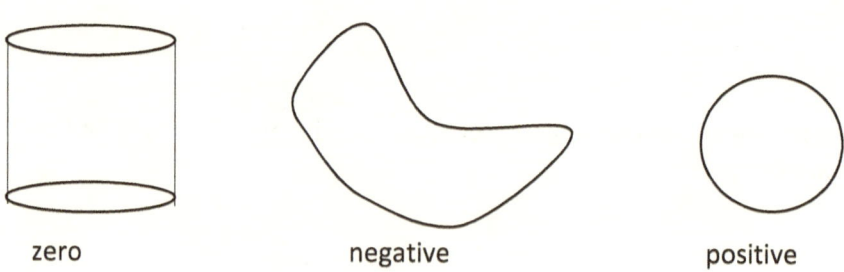

zero   negative   positive

But we live in a four-dimensional world with three space and one-time direction. So Gaussian curvature is not enough to describe it. The Riemannian curvature deals with space time geometry and is obviously much more tedious to deal with. We will explore it later.

Why do we bother about positive and negative curvature?

The reason is geometry is different. The flat surface geometry is called Euclidean geometry. This is what we study in school.

In Euclidean geometry, sum of angles in a triangle add up to 180 degree. The Pythagoras theorem applies to Euclidean geometry. The parallel lines stay parallel and so on.

The curved space geometry is called the **non-Euclidian** geometry. In case of positive curvature, sum of angles of a triangle is more than 180 degree and parallel lines converge. On the other hand, negative curvature causes the sum of angles in a triangle to be less than 180 degree with lines that are divergent or moving away from each other.

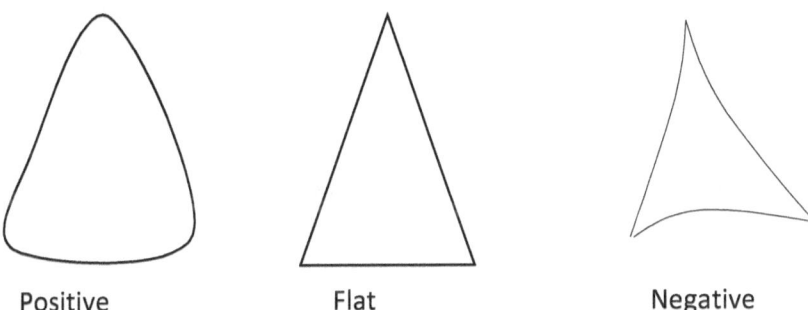

Positive   Flat   Negative

How do we draw straight lines on a curved surface?

We can go by the Euclidian definition of straight line, which is the shortest distance between two points. If the surface is curved, the shortest distance will not look like a straight line that we are accustomed too on a flat surface. But this is the best what can be done. The straightest possible lines on a curved surface are called **geodesics**.

On earth, shortest distance between two points lies on great circles. A great circle is any plane that passes through the center of the sphere e.g. equator is a great circle. If you slice the equatorial plane fully going through the center of the earth, then you will cut earth into two halves. If you have to travel between two cities that lie on the equator, just follow the equator, this is the shortest route.

A quick review of some technical terms that you may encounter in your quest to learn General Relativity is desirable. Technical term used for space is a **manifold**. It has a precise meaning but loosely speaking, it is formed by a topological space that looks flat at a local level. This is not surprising, we live locally on earth, it surely appears flat to us! This is an important fact and will be used to develop equations of General Relativity. It also means local inertial frames (flat coordinates) are always possible but there is no global inertial frame. A topological space is also a technical term, it is made of set of points that share certain relationships. The sets can be open or closed based on open and closed boundaries. The topological space can be flexible with various shapes sharing same topology e.g. topology of line and circle is same. A manifold can have various dimensions. The dimensions are usually given by real positive integers e.g. 1,2,3 etc. The notation for those manifolds is R^n. The n represents the dimensionality of the manifold. R^1 manifold contains curves. R^2 manifold contains surfaces. Sometimes you will see things written like surfaces in R^3. This just means that you are looking at surface of a 3-dimensional sphere. The manifolds used in physics are differentiable. We have seen tangents used to defined curvature. They play a very important role. Every point on a manifold has a tangent, that sticks out a vector which lives in a tangent plane, not in the manifold but just tangent to it. The collection of all tangents of a manifold is called tangent bundle. The tangents will be used to develop formula for geodesics in later chapters.

What is space time curvature?

Hopefully you have got some idea about curvature of space. Now we need to add time to the dimensions. How can time be bent or curved? I know it goes against our intuition that time is just a click on the clock, there is no way it can be curved. But as we have seen in Special Relativity, time is relative. Moving clocks run slow. This is taken to a whole new level in General Relativity.

In relativity, proper distance and time, which is measured at rest with respect to an event is paramount. This is because all observers agree on it, in technical words, proper distance and time are **Lorentz invariant**.

We already studied proper distance which is defined as

$$S^2 = c^2t^2 - x^2 - y^2 - z^2$$

To get proper time, just divide by c^2 to convert to time units.

$$\tau^2 = t^2 - \frac{x^2}{c^2} - \frac{y^2}{c^2} - \frac{z^2}{c^2}$$

Most books set c=1 and make life easier. I will try to do that as well. In some books, you will find proper time and distance signs changed. It's just a convention. Important point is that time and spatial distances carry opposite signs.

$$S^2 = -c^2t^2 + x^2 + y^2 + z^2$$
is equally valid.

In special relativity, meter sticks may measure different values, but sticks are measuring on a flat surface which is straight forward. But how do we use meter sticks on a curved surface? Try measuring distance between two points on a tennis ball with a ruler. The ruler will stick out of the ball. You have to get a flexible measuring tape that can be bent to measure distances on a curved surface.

You may say that it is obvious, earth is curved but what about the distance between earth and moon? There is empty space between them. Surely that cannot be curved. This is where our intuition fails us. Empty space is not just a passive background. It's a dynamical space where General Relativity predicts curvature of space time of even empty space! So flat coordinates cannot be used to describe a curved space. That's the very definition of curved space. We already discussed, if you can flatten a thing on to a flat surface meaning flat coordinates can be used to describe a thing, then it cannot be curved. In a curved space time, everything is curved, empty space including meter sticks that are taken there to measure it.

Time is a dimension in General Relativity. In flat space time, we can use a time meter stick with time clicks as units.

Time stick-click, click, click.

But what if time is curved; how do we measure it?

The only way is to find proper time, like a clock attached to the curved time dimension. Obviously, it is going to have more clicks as time dimension is curved, not straight so there are more clicks on it. In other words, **time slows down in a curved space time**. This has profound implications as we will see when we study black holes where there is extreme curvature of space time.

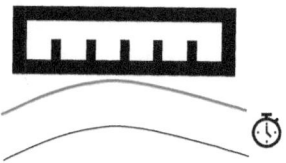

Time and space coordinates that we use, does not mean this is what is measured in experiments. This is a tricky point to remember. Only proper distance and time have significance. The coordinates can always be changed.

Let's end this chapter on a lighter note. What's the shortest distance or geodesic to get elected to Congress?

This is easy, be a part of a political dynasty through relationships, go to law school, then the geodesic to political success is a straight line. For the rest, geodesics continue to remain elusive!

Chapter 5

Space-Time Coordinates

It's time to put coordinates on our flat and curved spaces. The simplest coordinates are meter sticks. If I tell you that a table is 10 m away from me in x direction, that's one-dimensional coordinate. We can use x-y plane to form two-dimensional system or go in three dimensions to see the Cartesian coordinates in all their glory.

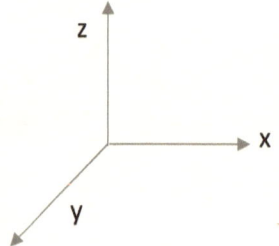

The three directions of coordinates are at 90 degree to each other. The technical name to describe it is they are orthogonal. There is no overlap between the coordinates. A coordinate is made up of several basis vectors. The simplest basis vector is the unit vector with magnitude of 1. It simply means that we have divided the direction into say 1-meter sticks. The unit vectors are sometimes denoted by a hat over them, $\hat{x}$ unit vector. Another name given to x, y and z unit vectors is i, j, k vectors. We could write that a table is $10\hat{x}$ or $10i$ away from the origin.

The formula for finding magnitude for any line segment is basically Pythagoras theorem.

$l^2 = \Delta x^2 + \Delta y^2 + \Delta z^2$

Where Δ (delta) is the difference in coordinates from one point to another on the line segment. I will skip writing delta in subsequent discussion to keep equations clean.

If line segment is a curve, then we have to use calculus.

$dl^2 = dx^2 + dy^2 + dz^2$

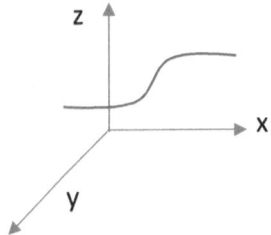

The curve length can be found by integrating and parametrizing the curve. The parameter will be the integration variable. Parameterization simply means using a parameter that is useful. If most of the curve is in x direction, then we can use x as the parameter. For a politician, elections are the parameter, no matter what the issue is!

$l = \int \sqrt{dx^2 + dy^2 + dz^2} \, du$ where u is the parameter of the curve.

Matrix notation is a useful way to represent coordinates.

We arbitrarily choose $x = \begin{pmatrix} 1 \\ 0 \\ 0 \end{pmatrix}$, $y = \begin{pmatrix} 0 \\ 1 \\ 0 \end{pmatrix}$ and $z = \begin{pmatrix} 0 \\ 0 \\ 1 \end{pmatrix}$

If we have a vector $l = 5i + 5j + 5k$ then in matrix notation, it will be

$l = (555)$

In z direction, the value of vector will be $(555)\begin{pmatrix}0\\0\\1\end{pmatrix} = \begin{pmatrix}0\\0\\5\end{pmatrix}$ or 5k. This is obvious but I wanted to illustrate the matrix notation. We could join x, y and z matrices into a single matrix. Here I have written basis vectors as row vectors (left to right).

$$\begin{pmatrix}1 & 0 & 0\\0 & 1 & 0\\0 & 0 & 1\end{pmatrix}$$

This type of matrix is a diagonal matrix as only diagonal has non zero values. Off diagonal terms are zero meaning there are no xy or yz terms. This means no overlap in coordinates. The matrix of coordinates is also called a **metric**.

A metric represents relationships between coordinates, like is there any overlap, how are they related, are they fixed or change with time, what's the spread of coordinates etc.

Overlap means dot product of coordinates. How much a coordinate projects onto another is the dot product.

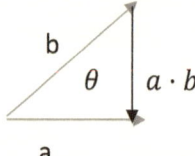

Dot product = $a \cdot b = ab\cos\theta$

Since cartesian coordinates are at 90 degree to one another, so dot product is zero between them. The metric of flat three-dimensional space in Cartesian coordinates can be written as

$$\begin{pmatrix}x\cdot x & x\cdot y & x\cdot z\\y\cdot x & y\cdot y & y\cdot z\\z\cdot x & z\cdot y & z\cdot z\end{pmatrix} = \begin{pmatrix}1 & 0 & 0\\0 & 1 & 0\\0 & 0 & 1\end{pmatrix}$$

Note off diagonal terms like $x \cdot y$ and $y \cdot x$ are same when calculating inner product. This type of matrix is called **symmetrical** matrix. This greatly reduces the number of calculations that are needed.

Let's look at the line segment or proper distance in special relativity.
$$S^2 = c^2 t^2 - x^2 - y^2 - z^2$$

Can you guess the metric for special relativity? The space for special relativity is called the **Minkowski** space. It represents flat space time.

The answer is straightforward. The unit vectors are same as three-dimensional space except the sign difference between time and space components. It is called the **signature** of the metric. In this case, signature is (+ - - -). The traditional symbol for the metric of special relativity is η (eta).

$$\eta = \begin{pmatrix} 1 & 0 & 0 & 0 \\ 0 & -1 & 0 & 0 \\ 0 & 0 & -1 & 0 \\ 0 & 0 & 0 & -1 \end{pmatrix}$$

The metric is diagonal and symmetric, same as three-dimensional space.

There is always a question of which coordinate frame be taken as reference. People in Toronto believe they are the center of universe and it should be ground zero for all coordinates. New Yorkers would obviously think differently. The frames can be moved or translated in different directions. They can be rotated in each direction as well. The coordinate frame may be moving with velocity in a direction which is called a boost. All these transformations have formulae that can be arranged in a matrix. The transformations are collectively called Poincare group. The boosts and rotations are non-commutable. This means if you rotate from one to another point and then reverse the process, you do not get back to the original point. These types of transformations are part of the so called non-Abelian group. On the other hand, if going from point a to b and then from b to a is same as seen in translations, they form Abelian group.

Polar coordinates

The Cartesian coordinates are not the only show in flat town. In a two-dimensional plane, distance and angle from origin can replace x and y coordinates.

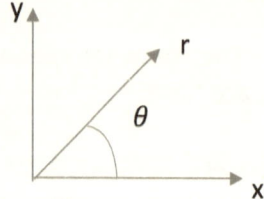

The relationship between Cartesian coordinates and Polar coordinates can be calculated by using basic trigonometry.

$x = r \cos \theta$
$y = r \sin \theta$

The line segment in polar coordinates can be calculated after differentiation of the above relations.

$dl^2 = dr^2 + r^2 d\theta^2$

The metric is $\begin{pmatrix} 1 & 0 \\ 0 & r^2 \end{pmatrix}$

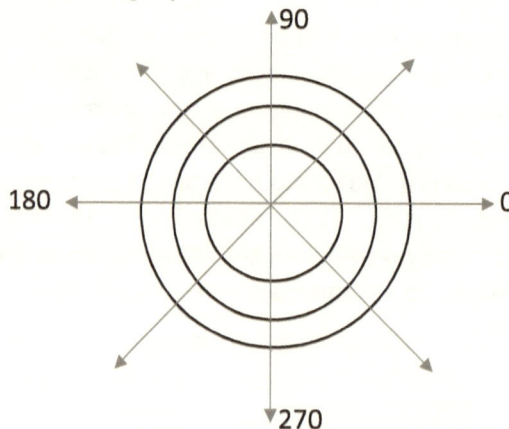

The constant values of r lie on circles and constant angles lie on arrows. As you can see, distance between angle coordinates increases as we move away from the origin. This is represented by r^2 term in the metric.

Polar coordinates are a convenient way of representing flat space. We can go back and forth between Cartesian and Polar coordinates as both represent flat space.

Spherical Coordinates

We move on to three-dimensional space because let's get real here! It's an important coordinate system as it is useful to describe a sphere which may be a hydrogen atom or earth or even a black hole. We have to add an angle that rotates around the z axis to get to spherical coordinates from polar coordinates.

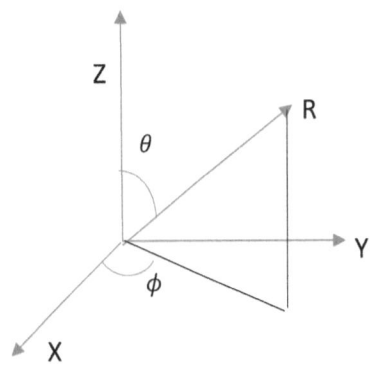

$x = r \sin\theta \cos\phi$
$y = r \sin\theta \sin\phi$
$z = r \cos\theta$

The line segment is

$dl^2 = dr^2 + r^2 d\theta^2 + r^2 \sin^2\theta d\phi^2$

The metric is

$$\begin{pmatrix} 1 & 0 & 0 \\ 0 & r^2 & 0 \\ 0 & 0 & r^2\sin^2\theta \end{pmatrix}$$

The spherical coordinates have curved intrinsic geometry which means they have non-Euclidean properties like sum of angles of triangle is not 180 degree and circumference of circle deviates from $2\pi r$.

Something to note here. There is no time dependence of coordinates. They will look the same even after billions of years. There are no cross terms like $d\phi\, d\theta$. If there are cross terms, then off diagonal terms will be non-zero and things will get a lot more complicated.

Chapter 6

Relativistic Notation

Let me introduce the dreaded relativity notation. Half the battle in understanding relativity is to get a grip on the notation. It's a language on to itself. There are many indices floating around. They appear and disappear at will and if you do not have an intuition about the indices, you will get lost quickly in relativity.

We start with the very basics. How to describe a vector?
It has magnitude and direction e.g. a bank is 100m west of the highway. Mathematically we could describe a vector as

$$V = V^i e_i$$

V^i is the magnitude and e_i represents the basis vector of the directional coordinate.

In relativity, spatial and time coordinates are denoted by

Time= x^0
Space coordinates=x, y and z = x^1, x^2 and x^3

A vector may have coordinates in all directions

$$V = \sum_{i=0}^{3} V^i e_i$$

The index i takes values of the four space-time coordinates.

In full glory, it means

$$V = V^0 e_0 + V^1 e_1 + V^2 e_2 + V^3 e_3$$

There is no dearth of summations in relativity. To avoid writing them over and over again, Einstein came up with a clever idea. He said if there are same upper and lower indices written together, it automatically means summation over all the indices.

It is called **Einstein summation** convention. Only a physicist could come up with such a convention, mathematicians would cringe at such conventions. The whole fun for mathematicians is writing fancy symbols over and over again. Physicists look past the redundant symbols and take a pragmatic view.

If you see $V^i e_i$, the sum over all the indices is implied.

You may ask why to write $V^i e_i$, why not $V_i e_i$?

It is a convention like many things in physics. It looks elegant and visually appealing.

Let me introduce you to another beast, $V_i e^i$. It is a dual vector. If you know quantum mechanics, you are aware of dual vectors which are complex conjugates of vectors. They are like mirror image of the vector. I do not want to confuse you as quantum mechanics is a much more abstract theory. Let's analyze these vectors in an intuitive way.

To make things simpler, I am going to just drop basis vector notation. Instead of $V^i e_i$, I will just write V^i. We will call it a vector with upper indices.

The vectors with upper indices are called **contravariant vectors**. The vectors with lower indices are called **covariant vectors**.

The covariant vectors are the perpendicular projections onto the coordinate axis. The contravariant vectors are the projections along the coordinate axis. Both mean the same thing for Cartesian coordinates as coordinates are perpendicular to one another. This is not the case with inclined coordinates.

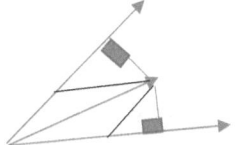

As you can see, the parallel projections(contravariant) are different than perpendicular(covariant) projections.

Why are they called contravariant and covariant?

Contravariant vector means, it goes in opposite direction to the change in basis vectors. If basis vectors are made bigger, the magnitude of the contravariant component of the vector decreases and vice versa. Think of contravariant vectors as measuring sticks. If an object is 10m, then decreasing m to cm, causes the length to become 1000 cm. So, you can see the contravariant component of the vector becomes bigger as the basis vector becomes smaller.

Covariant vector goes in the same direction as change in the basis vectors. A good example is a gradient. Going downhill or moving from higher density to lower density are examples of a gradient.

The vertical lines are basis gradient vectors. To represent a thick area of an object e.g. a rock, more vertical lines are needed. Greater the number of lines, greater is the value of thickness of the object and vice versa. So, the covariant component of the vector goes along with the change of basis vector. In other words, it covaries.

A vector could have contravariant and covariant components at the same time. This is obvious. We can always project parallel and perpendicular projections for a vector. But the important point to learn is how to multiply contravariant and covariant components.

$$V = V^i V_i$$

Note V has no indices left, they ate each other! Physicists don't use such strong language. The technical term is the **contraction of indices** has produced a scalar.

Summation is implied, meaning

$$V = V^0 V_0 + V^1 V_1 + V^2 V_2 + V^3 V_3$$

The indices are multiplied and summed over.

Does the above formula remind you of something?

Of course, the line segment

$$l^2 = x^2 + y^2 + z^2$$

In Cartesian coordinates, there is no difference between contravariant and covariant vectors as coordinates are perpendicular to one another. So, writing indices for the line segment is unnecessary.

But line segment or proper distance is an invariant term. It means no matter what coordinates we use; line segment or proper distance does not change. It is like measuring a car. You can use meters or cm to measure, it does not change length of the car. This is what scalar V represents. It is like the length of the car which is invariant to coordinate change.

Another thing to remember is that contraction streamlines the information contained in indices. The line segment gives us a single number e.g. 5m long car. In the process, information contained in indices like length of each coordinate direction is hidden away. This will have repercussions as we use contraction of indices to develop Einstein field equations.

Let's put indices on the metric. The traditional symbol for metric is g. It is given two indices $\mu\nu$. Both indices run from 0 to 3 covering one time and three spatial components. $g^{\mu\nu}$ is a metric with its dual given by $g_{\mu\nu}$. g^{11} is the space-space component denoting x.x dot product meaning overlap of the components. It forms the diagonal component of the matrix. g^{12} is the x.y component of the matrix and in case of Cartesian coordinates or flat Minkowski space, its zero as there is no overlap between coordinates. It forms the off-diagonal component of the metric. This game is played for all components to form the matrix.

$$\eta^{\mu\nu} \text{ or } g^{\mu\nu} = \begin{pmatrix} g^{00} & g^{01} & g^{02} & g^{03} \\ g^{10} & g^{11} & g^{12} & g^{13} \\ g^{20} & g^{21} & g^{22} & g^{23} \\ g^{30} & g^{31} & g^{32} & g^{33} \end{pmatrix} = \begin{pmatrix} 1 & 0 & 0 & 0 \\ 0 & -1 & 0 & 0 \\ 0 & 0 & -1 & 0 \\ 0 & 0 & 0 & -1 \end{pmatrix}$$

$\eta^{\mu\nu}$ (eta) is the traditional name for the metric of Minkowski space.

The dual metric of Minkowski space has the same matrix components.

$$\eta_{\mu\nu} \text{ or } g_{\mu\nu} = \begin{pmatrix} g_{00} & g_{01} & g_{02} & g_{03} \\ g_{10} & g_{11} & g_{12} & g_{13} \\ g_{20} & g_{21} & g_{22} & g_{23} \\ g_{30} & g_{31} & g_{32} & g_{33} \end{pmatrix} = \begin{pmatrix} 1 & 0 & 0 & 0 \\ 0 & -1 & 0 & 0 \\ 0 & 0 & -1 & 0 \\ 0 & 0 & 0 & -1 \end{pmatrix}$$

The metric is used to convert upper indices into lower indices vectors and vice versa.

$$V^{\mu} g_{\mu\nu} = V_{\nu}$$

$$V_{\mu} g^{\mu\nu} = V^{\nu}$$

There is no magic happening here. We are doing contraction of indices which is basically matrix multiplication.

To see the contraction of indices in action, let's define vectors first, based on convention.

$$V^\mu = \begin{pmatrix} ct \\ x \\ y \\ z \end{pmatrix}$$

$$V_\mu = \begin{pmatrix} ct \\ -x \\ -y \\ -z \end{pmatrix}$$

$V^\mu g_{\mu\nu} = V_\nu$ means

$$\begin{pmatrix} ct \\ x \\ y \\ z \end{pmatrix} \begin{pmatrix} 1 & 0 & 0 & 0 \\ 0 & -1 & 0 & 0 \\ 0 & 0 & -1 & 0 \\ 0 & 0 & 0 & -1 \end{pmatrix} = \begin{pmatrix} ct \\ -x \\ -y \\ -z \end{pmatrix}$$

We have multiplied and summed over indices; this is contraction of indices.

$ct \times 1 + x \times 0 + y \times 0 + z \times 0 = ct$

$ct \times 0 + x \times -1 + y \times 0 + z \times 0 = -x$

$ct \times 0 + x \times 0 + y \times -1 + z \times 0 = -y$

$ct \times 0 + x \times 0 + y \times 0 + z \times -1 = -z$

What is $V^\mu V_\mu$?

That's obvious, it is scalar V, which is line segment.

$V^\mu V_\mu = c^2 t^2 - x^2 - y^2 - z^2$ is the space-time distance called s^2 and is a Lorentz invariant quantity.

$V^\mu V^\mu = c^2t^2 + x^2 + y^2 + z^2$ is not Lorentz invariant term and is thus not used.

In the differential form, proper distance can be written as

$$ds^2 = g^{\mu\nu} dx_\mu dx_\nu$$

It means the same matrix multiplication process as above. It is a neat way of writing indices. We see again two indices being contracted to form an index free invariant scalar.

First when you multiply $g^{\mu\nu} dx_\mu$, this is same as

$$\begin{pmatrix} ct \\ -x \\ -y \\ -z \end{pmatrix} \begin{pmatrix} 1 & 0 & 0 & 0 \\ 0 & -1 & 0 & 0 \\ 0 & 0 & -1 & 0 \\ 0 & 0 & 0 & -1 \end{pmatrix} = \begin{pmatrix} ct \\ x \\ y \\ z \end{pmatrix}$$

Then multiplying by dx_ν means

$$\begin{pmatrix} ct \\ -x \\ -y \\ -z \end{pmatrix} \begin{pmatrix} ct \\ x \\ y \\ z \end{pmatrix} = c^2t^2 - x^2 - y^2 - z^2 = ds^2$$

Differentiation with respect to space-time coordinates is written in a short hand notation.

$\frac{\partial}{\partial x} = \partial^x$ and $-\frac{\partial}{\partial x} = \partial_x$

So, $\partial^\mu \partial_\mu = \frac{\partial}{c^2 \partial t^2} - \frac{\partial}{\partial x^2} - \frac{\partial}{\partial y^2} - \frac{\partial}{\partial z^2}$

The four space time coordinates in special relativity form **four vectors**.

The position vector, we already know

$$x^\mu = x^0, x^1, x^2, x^3$$

Next is the Four-velocity, U^μ

It is formed by differentiating four-position vector with respect to proper time, τ

$$\frac{dx^\mu}{d\tau} = \frac{c\,dt}{d\tau}, \frac{dx}{d\tau}, \frac{dy}{d\tau}, \frac{dz}{d\tau}$$

In terms of velocity in a frame, it is

$U^\mu = U^0, U^1 U^2, U^3 = c\gamma, \gamma v_x, \gamma v_y, \gamma v_z$

The four momentum is straight forward

$P^\mu = mU^\mu$

We already derived its expression earlier, so plugging in the value leads to
$P^\mu = mU^\mu = \gamma mc, \gamma mv(x, y, z)$

The time component of momentum or P^0 can be written as $\frac{E}{c}$.

It is useful to form invariant quantities, that are true in all reference frames.

So, in the spirit of $ds^2 = g^{\mu\nu} dx_\mu dx_\nu$

We can form a similar invariant quantity

$\eta_{uv} U^\mu U^\nu$

We have to put value of U^μ calculated earlier and do the calculation. It will be something like $\gamma^2 c^2 - \gamma^2 v^2$.

This will simplify to c^2.

So, we have $\eta_{uv} U^\mu U^\nu = c^2$

This equation will be used to simplify Einstein field equations when doing certain calculations.

How do we change coordinates of contravariant and covariant vectors?

The change of coordinates is the central theme of relativity. We need to develop clear transformational properties of vectors. The contravariant and covariant vectors transform differently under change of coordinates.

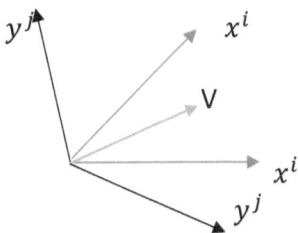

We want to change from x^i ($i = 0,1,2,3$) to y^j ($j = 0,1,2,3$) coordinate system. The coordinates could be flat or curved, does not matter.

We start with how a small change in x coordinates leads to small change in y coordinates. So, we need to do differentiation.

$dy^j = \frac{\partial y^j}{\partial x^i} dx^i$

In terms of the total change of vector components,

$V^j = \frac{\partial y^j}{\partial x^i} V^i$

This is the transformational property of contravariant vectors.

How does it square with the intuition of contravariant vectors as meter sticks?

Again, let's do a trivial transformation from meter to km, say 1000m to 1 km.

$$V^j = \frac{\partial y^j}{\partial x^i} V^i \text{ or } 1(km) = \left(\frac{1}{1000}\right) 1000(m)$$

If you look carefully, there is i index upstairs and downstairs. This means summation of i index (0 to 3). j index is not summated.

$$V^0 = \frac{\partial y^0}{\partial x^0} V^0, V^0 = \frac{\partial y^0}{\partial x^1} V^1, V^0 = \frac{\partial y^0}{\partial x^2} V^2, V^0 = \frac{\partial y^0}{\partial x^3} V^3$$

The transformational property of covariant vector is

$$\frac{\partial \rho}{\partial y^j} = \frac{\partial x^i}{\partial y^j} \frac{\partial \rho}{\partial x^i} \text{ or}$$

$$V_j = \frac{\partial x^i}{\partial y^j} V_i$$

The differentiation is with respect to differential elements with upper indices, that's ok. i index is again summated.

Covariant vectors represent gradient with units of $\frac{1}{distance}$.

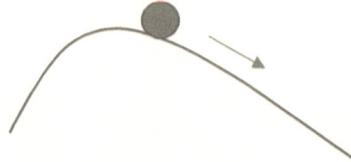

Think of a ball sliding down a hill(gradient). The ball gains momentum or force as it goes down the gradient.

$$\frac{1}{m} \sim F$$

Then going from m to km leads to

$$\frac{1}{\frac{1}{1000}km} \sim 1000F$$

So, it covaries with the change of coordinates.

The objects that obey transformation properties like above are called **Tensors**. We can map tensors to other tensors through transformation properties of coordinates.

We can combine tensors.

$$V^j P^r = \frac{\partial y^j}{\partial x^i} V^i \frac{\partial y^r}{\partial x^s} P^s \text{ or } \frac{\partial y^j}{\partial x^i} \frac{\partial y^r}{\partial x^s} V^i P^s$$

$$T^{jr} = \frac{\partial y^j}{\partial x^i} \frac{\partial y^r}{\partial x^s} T^{is}$$

T^{jr} is a rank 2 tensor as it has two indices. We have converted a two-rank tensor into another. Scalar function is a rank zero tensor, vector is a rank one tensor.

Similarly, lower indices can be combined to form covariant rank two tensor.

$$T_{jr} = \frac{\partial x^j}{\partial y^i} \frac{\partial x^r}{\partial y^s} T_{is}$$

A mixed lower and upper index tensor can be formed as well.

$$T_r^j = \frac{\partial x^r}{\partial y^s} \frac{\partial y^j}{\partial x^i} T_s^i$$

We can form higher rank tensors via multiplication in the same manner.

We have seen multiplication and contraction of tensors. But usual algebraic operations like addition and subtraction can be done as well.

$$T_j^i + S_j^i = (T + S)_j^i = W_j^i$$

Only tensors of the same contra and covariant ranks can be added or subtracted. The addition and subtraction take place component by component.

$T_j^{ir} + S_j^i$ is not allowed.

What about the metric?

You guessed it right, it is a tensor. In fact, it is the most important tensor in relativity. Let's see its transformational properties.

$$ds^2 = g_{\mu\nu}dx^\mu dx^\nu$$

Substituting the transformational properties of contravariant vectors,

$$dx^\mu = \frac{\partial x^\mu}{\partial y^i}dx^i, \quad dx^\nu = \frac{\partial x^\nu}{\partial y^j}dx^j$$

$$ds^2 = g_{\mu\nu}\frac{\partial x^\mu}{\partial y^i}\frac{\partial x^\nu}{\partial y^j}dx^i dx^j$$

$$ds^2 = g_{ij}dx^i dx^j$$

Where $g_{ij} = g_{\mu\nu}\frac{\partial x^\mu}{\partial y^i}\frac{\partial x^\nu}{\partial y^j}$

The matrix components of the metric tensor are transformed by the differentials as shown above to get the new metric tensor. The matrix formed by the differentials is called the Jacobian matrix.

The upper and lower index metric tensors are inverse of each other.

If the metric is $g_{ij} = \begin{pmatrix} 1 & 0 \\ 0 & r^2 \end{pmatrix}$ which is the case with polar coordinates, then the dual metric g^{ij} will be $\begin{pmatrix} 1 & 0 \\ 0 & \frac{1}{r^2} \end{pmatrix}$

The multiplication of both matrices is an identity matrix which is basically 1, a scalar as both indices are contracted.

$$g_{ij}g^{ij} = I = \begin{pmatrix} 1 & 0 \\ 0 & 1 \end{pmatrix}$$

Differentiation of tensors

Differentiation basically means how a vector changes from place to place.

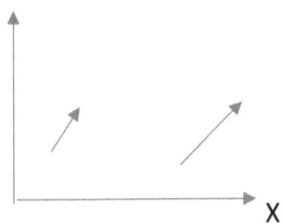

As vector is moved along the x axis, change is $\frac{\partial v}{\partial x}$. This assumes that coordinate x is not changing from one place to another.

Differentiation of tensors is tricky. If a vector does not change from one place to another in a coordinate frame, it does not mean that it will stay the same in another coordinate frame. So, ordinary differentiation is not a tensor operation.

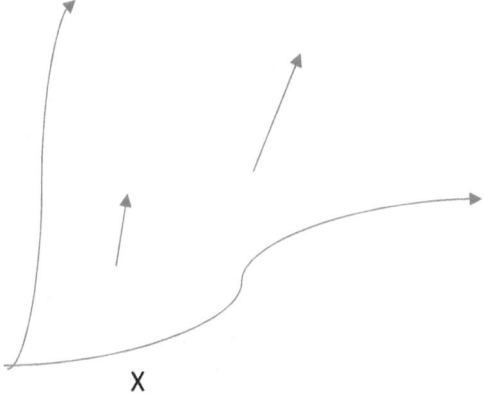

If coordinates change from one place to another, that has to be tracked as well.

A tensor differentiation involves differentiation of vector plus differentiation of change of coordinates. It is called **covariant differentiation**. Don't get confused with the terminology. Covariant differentiation does not mean it is only meant for covariant vectors. In this context, it is used generically which means it applies to both covariant and contravariant vectors. There is no such thing as contravariant differentiation!

$$\nabla_i V^j = \frac{\partial V^j}{\partial x^i} + \Gamma^j_{ki} V^k$$

∇_i is the symbol for covariant differentiation.

$\frac{\partial V^j}{\partial x^i}$ is just the ordinary differentiation of the vector V^j.

$\Gamma^j_{ki} V^k$ represents the change of coordinates. Note k index is summed over.

Γ is called a connection coefficient or **Christoffel symbol**.

e.g. if we want to differentiate with respect to time

$\nabla_t V^x = \frac{\partial V^x}{\partial x^t} + \Gamma^x_{kt} V^k$ and fill in the value of k that goes from 0 to 3 again, we will form four equations.

Play the same game with all the variables like $\nabla_y V^x$ and then keep summing over all the indices for Γ. These formulae are an equation producing factory, they add up quickly!

What do the indices on Christoffel symbol represent?

Think of like this, Γ^x_{yz} means how x coordinate changes if y changes with respect to z. If y is summed over then you have to include all the combinations like how x coordinate changes if z changes with respect to y and so on.

Γ is not a tensor as its zero in Cartesian coordinates as x, y and z in that case are 90 degrees to one another, so there is no overlap. This means x is not affected by change of y and z coordinates. But Γ is not zero for curved coordinates. The very definition of a tensor is that its transformational properties do not change from coordinate to coordinate. Γ clearly does not obey this condition.

The covariant differentiation of lower index or covariant vectors is similar to the upper index or contravariant vectors expect for minus sign.

$$\nabla_i V_j = \frac{\partial V^j}{\partial x^i} - \Gamma_{ji}^k V_k$$

If you are still not clear about this fancy mathematics, then I am going to give a simple analogy. Think of a car speeding on a motor way. Its speed will depend on the car and the driver, that's vector differentiation. But that's not all! The road conditions do matter. A bumpy road, snow or ice on the road is obviously going to affect speed of the car as well. This is represented by Γ. That's why when we are comparing lap times of racing drivers, its useful to compare whether it was a wet or a dry lap.

We can extend the covariant differentiation to mixed and higher rank tensors.

$$\nabla_i T^{rj} = \frac{\partial T^{rj}}{\partial x^i} + \Gamma_{ki}^r T^{kj} + \Gamma_{ki}^j T^{rk}$$

And for mixed tensors

$$\nabla_i T_j^r = \frac{\partial T_j^r}{\partial x^i} + \Gamma_{ki}^r T_j^k - \Gamma_{ji}^k T_k^r$$

Index k is summed over, so for each combination of indices (0 to 3), you have to sum over all the combinations. This means there are lots of equations to calculate. That's why doing calculations in relativity is a messy process.

How do we calculate Γ ?

We know that it represents change in coordinates. Since overlap or relationship between coordinates is determined by the metric tensor, we should expect that Γ should also depend on metric tensor, and it does.

We will use the fact that we learned earlier that no matter how curved a surface is, locally we can always use flat coordinates. It is like saying that pick any curved object, you always place a dot with a pen anywhere on it. This is analogous to choosing local flat coordinates on a curved space.

The mathematical way of saying it is that the metric tensor does not change locally, which means the first derivative of metric tensor is zero. The second derivative, which means change in the rate of change of metric cannot be zero as eventually the space does curve.

It also means that covariant derivative of metric tensor is zero. This is obvious and trivial. Since metric tensor is not changing locally, it is true for all coordinate frames, that's the definition of covariant differentiation.

$$\nabla g_{\mu\nu} = 0$$

Putting in the formula for covariant derivative

$$\nabla_i g_{\mu\nu} = \frac{\partial g_{\mu\nu}}{\partial x^i} - \Gamma^r_{\nu i} g_{\mu r} - \Gamma^r_{\mu i} g_{r\nu}$$

To isolate only Γ, we have to make three equations with different indices like above and add, subtract them. I won't go into the detailed derivation as its couple of pages long, but concept is straight forward. There will be some cancellations and we will be able to isolate Γ.

After algebraic calculations we will be left with

$$g_{rk}\Gamma^r_{\mu i} = \frac{1}{2}(\frac{\partial g_{ki}}{\partial x^\mu} + \frac{\partial g_{\mu k}}{\partial x^i} - \frac{\partial g_{\mu i}}{\partial x^k})$$

To take g_{rk} to the other side, we replace it with the dual metric which was introduced previously.

$$\Gamma^r_{\mu i} = g^{rk}\frac{1}{2}(\frac{\partial g_{ki}}{\partial x^\mu} + \frac{\partial g_{\mu k}}{\partial x^i} - \frac{\partial g_{\mu i}}{\partial x^k})$$

Index k is summed over.

Don't fret over index placements and trying to memorize the formulae, focus on the meaning and intuition behind metric and connection coefficients.

Let's do some examples to see how the machinery of connection coefficients work.

First, let's take the example of flat space and Cartesian coordinates of three-dimensional space.

The metric and inverse metric is obviously same in this case as inverse of 1 is 1.

$$g_{\mu\nu} = g^{\mu\nu} = \begin{pmatrix} 1 & 0 & 0 \\ 0 & 1 & 0 \\ 0 & 0 & 1 \end{pmatrix} = \begin{pmatrix} g_{11} & g_{12} & g_{13} \\ g_{21} & g_{22} & g_{23} \\ g_{31} & g_{32} & g_{33} \end{pmatrix} = \begin{pmatrix} g^{11} & g^{12} & g^{13} \\ g^{21} & g^{22} & g^{23} \\ g^{31} & g^{32} & g^{33} \end{pmatrix}$$

$$\Gamma^x_{yz} = g^{xk}\frac{1}{2}(\frac{\partial g_{kz}}{\partial x^y} + \frac{\partial g_{yk}}{\partial x^z} - \frac{\partial g_{yz}}{\partial x^k})$$

We could write ∂x^y as simply ∂y. Now we have to sum over k which means putting in value of k as 1,2 and 3.

Since we used number notation for metric, let's put x=1, y=2, z=3.

For k=1

$$\Gamma^1_{23} = g^{11}\frac{1}{2}(\frac{\partial g_{13}}{\partial x^1} + \frac{\partial g_{21}}{\partial x^3} - \frac{\partial g_{23}}{\partial x^1})$$

Put in the value of the metric components from above

$$\Gamma^1_{23} = 1 \cdot \frac{1}{2}(\frac{\partial\,0}{\partial x^1} + \frac{\partial\,0}{\partial x^3} - \frac{\partial\,0}{\partial x^1}) = 0$$

Differentiation of a constant like 0 is 0. Obviously if something is constant, how can it change!

You can put in all values of k; you will find the answer is always zero. You can try other combinations of connection coefficients like Γ^2_{13}, Γ^3_{21} etc., the answer is again zero. This is as expected as Cartesian coordinates do not change and there is no overlap between coordinates, so all connection coefficients are zero. Note due to symmetry, things like $\Gamma^3_{21} = \Gamma^3_{12}$, so number of calculations are reduced a bit.

Let's move on to curved coordinates where we expect to get non trivial connection coefficients. The simplest example is spherical coordinates. If you recall the line segment and metric for spherical coordinates is

The line segment is

$$dl^2 = dr^2 + r^2 d\theta^2 + r^2 \sin^2\theta d\phi^2$$

The metric is

$$g_{\mu\nu} = \begin{pmatrix} 1 & 0 & 0 \\ 0 & r^2 & 0 \\ 0 & 0 & r^2\sin^2\theta \end{pmatrix} = \begin{pmatrix} g_{11} & g_{12} & g_{13} \\ g_{21} & g_{22} & g_{23} \\ g_{31} & g_{32} & g_{33} \end{pmatrix}$$

$$g^{\mu\nu} = \begin{pmatrix} 1 & 0 & 0 \\ 0 & \frac{1}{r^2} & 0 \\ 0 & 0 & \frac{1}{r^2\sin^2\theta} \end{pmatrix} = \begin{pmatrix} g^{11} & g^{12} & g^{13} \\ g^{21} & g^{22} & g^{23} \\ g^{31} & g^{32} & g^{33} \end{pmatrix}$$

Here x ranges from $x^1 = r, x^2 = \theta$ and $x^3 = \phi$.

$$\Gamma^1_{23} = g^{11}\frac{1}{2}(\frac{\partial g_{13}}{\partial x^1} + \frac{\partial g_{21}}{\partial x^3} - \frac{\partial g_{23}}{\partial x^1}) = \Gamma^1_{23} = 1\frac{1}{2}(\frac{\partial 0}{\partial x^1} + \frac{\partial 0}{\partial x^3} - \frac{\partial 0}{\partial x^1}) = 0$$

Γ^1_{32} by symmetry is also zero. In fact, most of the connection coefficients are zero except a few.

$$\Gamma^2_{33} = g^{22}\frac{1}{2}(\frac{\partial g_{23}}{\partial x^3} + \frac{\partial g_{32}}{\partial x^3} - \frac{\partial g_{33}}{\partial x^2})$$

Let's assume it's a unit sphere so that we set value of $r = 1$. Then our connection coefficient will become

$$\frac{1}{(1)^2}\frac{1}{2}(\frac{\partial 0}{\partial x^3} + \frac{\partial 0}{\partial x^3} - \frac{\partial (1)^2 \sin^2\theta}{\partial x^2})$$

Simplifying, we have to basically differentiate $-\frac{1}{2}\frac{\partial \sin^2\theta}{\partial \theta}$.

This is basic differentiation; we will solve it in case you are rusty on the details

Let $u = \sin\theta$ so $\frac{\partial u^2}{\partial x} = 2u\frac{\partial u}{\partial x} = 2\sin\theta\frac{\partial \sin\theta}{\partial \theta} = 2\sin\theta\cos\theta$

$\Gamma^2_{33} = -\sin\theta\cos\theta$.

There you have it. If you have time, try to calculate all the connection coefficients.

What's the point of calculating Γ's?

As you will see, curvature and equations of motion have Γ's buried in them, so their calculation is critical to do any calculations in relativity.

Geodesics

Geodesics are the straightest possible lines between two points. There are no straight lines in a curved space, only straightest possible lines. A good way to prove geodesics is through tangent vector. We can draw a tangent at every point on the geodesic.

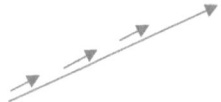

For a straight line in a flat space, tangent vectors, do not change, always parallel to the line.

This is not the case with a curved space.

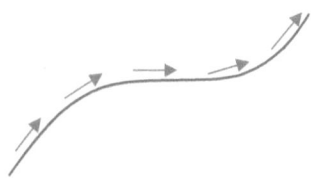

We need to do covariant differentiation. First let's take an arbitrary vector and move it along the tangent vector.

The tangent vector can be described as $\frac{dx^j}{dl}$.
It represents how coordinates vary along the line segment.

Covariant differentiation of a vector along tangent vector would be

$$\nabla_j V^i = \frac{\partial V^i}{\partial x^j}\frac{dx^j}{dl} + \Gamma^i_{jk} V^k \frac{dx^j}{dl} \text{ or}$$

$$\nabla_j V^i = \frac{\partial V^i}{\partial l} + \Gamma^i_{jk} V^k \frac{dx^j}{dl}$$

What if the vector to be transported along the tangent vector is tangent vector itself, then we will get

$$\frac{d^2 x^i}{dl^2} + \Gamma^i_{jk} \frac{dx^j}{dl}\frac{dx^k}{dl}$$

To be a geodesic, the condition is the tangent vector does not change in any coordinate system, that means $\nabla_j V^i = 0$

$$\frac{d^2x^i}{dl^2} + \Gamma^i_{jk} \frac{dx^j}{dl} \frac{dx^k}{dl} = 0$$

This is an important equation, called the **geodesic equation**.

It also means $\frac{d^2x^i}{dl^2} = - \Gamma^i_{jk} \frac{dx^j}{dl} \frac{dx^k}{dl}$

Enough with the equations, what does it imply?

It implies that change in the rate of change of tangent vector depends on how curvature or metric changes. That makes sense. Take a pen, try to keep it straight along any curved object like a sofa or a table. As you move along the bends of the table, keep the pen along the table as much as possible, this is what the geodesic equation implies, and it is the shortest distance in a curved space.

We know that instead of the proper distance or line segment, it could also represent proper time. If things are moving very slowly, then proper time is same as coordinate time.

$$d\tau^2 = dt^2 - \left(\frac{dx^2+dy^2+dz^2}{c^2}\right) \Rightarrow d\tau^2 = dt^2 \text{ as } \frac{dx^2+dy^2+dz^2}{c^2} \text{ is small}$$

$\frac{d^2x^i}{dt^2}$ term now represents ordinary acceleration of a particle.

$\Gamma^i_{jk} \frac{dx^j}{dl} \frac{dx^k}{dl}$ term needs some approximation. At slower speeds $\frac{dx^0}{dt} = 1$, other spatial velocity components by assumption are very small, let's ignore them.

We are left with Γ^i_{jk} only. Let's further assume the curvature is due to a stationary object like an earth so only time – time (x^0) component will be important.

Γ^1_{00}, Γ^2_{00}, Γ^3_{00} are relevant, lets pick one component and work with it.

$$\Gamma^1_{00} = g^{11} \frac{1}{2} (\frac{\partial g_{10}}{\partial x^0} + \frac{\partial g_{01}}{\partial x^0} - \frac{\partial g_{00}}{\partial x^1})$$

If the space is little bit curved or gravity is weak then metric is nearly flat space which means $\eta_{\mu\nu} + h_{\mu\nu}$ (small deviation from flat metric). We will work in first order in small corrections. It means we will put $h_{\mu\nu}$ only once in the equation.

We will take $g^{11} = \eta^{11}$ =- 1. No need to add correction here as working in first order only.

The metric is changing very slowly with time by assumption so $\frac{\partial g_{10}}{\partial x^0}$ and $\frac{\partial g_{01}}{\partial x^0}$ terms can be ignored as well.

We are left with $\frac{d^2 x^i}{dt^2} = -\frac{1}{2} \frac{\partial g_{00}}{\partial x^1}$

If we put in the correction, then $\frac{d^2 x^i}{dt^2} = -\frac{1}{2} \frac{\partial (\eta_{00} + h_{00})}{\partial x^1} = -\frac{1}{2} \frac{\partial (1 + h_{00})}{\partial x^1} = -\frac{1}{2} \frac{\partial h_{00}}{\partial x^1}$

From Newtonian gravity, we also know that the gravitation potential is related to acceleration due to gravity by

$$\frac{dx^2}{dt^2} = -\frac{d\phi}{dx}$$

So, we can relate gravitational potential to metric by integration, that will lead to

$h_{00} = 2\phi + constant$ or in terms of the whole metric

$g_{00} = 2\phi + 1$

I have not kept track of speed of light in these formulae for simplicity.

That's lot of assumptions we made! What's the point?

First of all, you see that the acceleration depends on the curvature of space. This is the starting point of general relativity. Second, by the tedious exercise above, we showed that in Newtonian limit, the geodesic equation resembles Newtonian gravity. This is very important. Whenever we have a new and a profound theory that is radically different than the previous well-established theories, it is critical that the new theory shows that in certain limits, it resembles the old tested theory. This keep physicists asleep at night!

Chapter 7

Einstein Field Equations

We are entering the climax of General Relativity. Whatever we have learned in previous chapters about curved coordinates and motion will be put into a compact equation which is crown jewel of relativity. But we will first start with basic concepts. This is how Einstein developed General Relativity. He had a strong intuition about certain physical principles. He did thought experiments and confirmed his intuition and came up with the implications of the theory. Mathematics was developed later. This is how a physical theory should be developed. That's why Einstein has otherworldly reputation about him. Very often we have experimental evidence and then try to build a theory to explain those experiments. It is not a satisfactory solution. Tweaking certain parameters to fit in experimental data does not reach the same level of satisfaction. Newer theories like Supersymmetry try to fit in the data this way and have not been successful thus far to knock down General Relativity or Quantum Mechanics from their pedestal. But we have to be realistic, how often do we get people like Newton and Einstein? We can aspire to be like them but realistically odds are against us.

Einstein started with a simple thought experiment. He thought what happens if a person falls freely under gravity?

A clarification, no one was hurt in this experiment!

A Newtonian physicist would say, a person falling freely under gravity is accelerating due to gravitational force given by Newton's law of gravitation.

Einstein had a profound and radical thought. He described it as the happiest moment of his life. He thought free fall means as if gravity has been switched off. That's why a person feels weightless in free fall. He argued that a person in space away from any gravitational object would feel the same way. Since free fall and person in gravity free space are inertial frames, they have same physical laws. This is the equivalence principle.

The things in reverse are also true. It means a person standing on earth feels weight due to gravity. This is analogous to a person in a spaceship accelerating upwards. His feet will feel the acceleration of the spaceship. There is no difference between this acceleration and gravity.

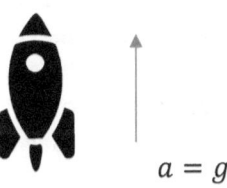

$a = g$

Due to equivalence principle, the effects of acceleration and gravity are identical. In other words, experiments done in the spaceship should extrapolate to gravity as well.

Suppose an astronaut shines light across the spaceship. What happens?

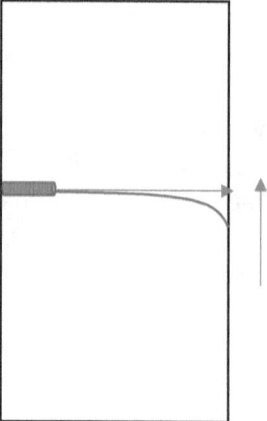

The astronaut will see the light strike straight across the wall. This is because spaceship is a small frame where local flat coordinates can be used. But this is not what a distant observer will see. A distant observer e.g. on earth, will see the light touch lower on the wall as by the time light left the torch, the wall has moved upwards as spaceship is accelerating upwards. In other words, he will see light bending. To explain this finding, distant observer will say that light was bent due to effect of gravity.

We have a remarkable conclusion, without writing any mathematics, with a simple thought experiment, we can conclude that gravity bends light. This is a critical experimental test to confirm relativity and subsequent experiments did validate this hypothesis.

Let's change the direction of torch to the ceiling of the spacecraft. What happens to the frequency of light?

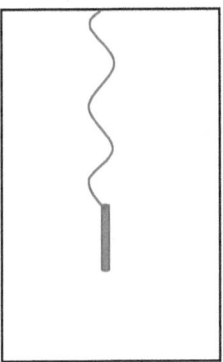

The person in the aircraft notices no difference in the frequency of light as it touches the ceiling. Again, this is due to the fact that spacecraft is small enough to use local flat coordinates. A distant observer will not see the same thing. He will see that the frequency of light decreases or **red-shifted**. This happens because as the light is going towards the ceiling, it is moving away from the light with a certain acceleration. It is as if light is chasing someone. Try chasing a runner, obviously you will get tired. In other words, your steps will get slower. In case of light, its energy or frequency decreases as it gets tired chasing the ceiling. The opposite is also true. If the ceiling was moving towards the light, the frequency of light as it reaches the ceiling will increase, this is called blue-shift. The color scheme of light is based on frequency of light, in case you forgot high school physics.

The change of frequency due to moving observers is nothing new. It is called the Doppler effect. We have all seen that an incoming car or train will cause the sound to be high pitched as it approaches us, and the pitch decreases as it recedes away from us. With relativity, this happens due to gravity as well. The light escaping gravity gets red shifted as it has to overcome the curvature of space time and doing so causes it to get tired and lose some of its frequency.

Einstein had these ideas and understanding of the consequences of these ideas like bending or red shifting of light due to gravity. But he did not know how to go about building mathematics around it. He had to learn about tensors and curvature from his friends and colleagues. It took about a decade from these thought experiments for Einstein to build his final theory or field equations.

We have already done the background work about curvature of space-time and relativistic notation. It is time to wrap things up and connect loose ends.

We can easily guess that any relativistic gravity equation has to be a tensor equation as only then it will transform from one coordinate system to another properly.

First, we have to build a tensor out of mass and energy which will act as a source of gravity. It's like saying, let's build a tensor out of earth or sun. It's not too difficult.

Where do we start?

Let's put earth in a box. The technical term is dust in a box.

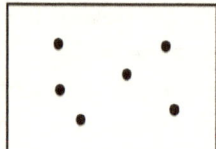

What's the density?

$$\rho = \frac{dD}{dV}$$

The density is how many dust particles are there in a volume. The usual volume is over three space coordinates (x, y and z).

The dust particles can be in motion as well. The motion is like a current.

$$J^x = \frac{dD}{dAdt}$$

The current is formed by flow over time. It measures how much area is covered per unit time. The y and z directions will also have their currents.

The dust that was stationary and is now moving should match each other to preserve energy conservation. It is like saying that water coming out of water tap should match its source which is the water tank. This is called the **continuity equation**.

$$\frac{\partial \rho}{\partial t} = -\nabla J$$

Minus sign simply means that as tap flows, water tank density decreases. Putting this equation into relativistic notation where time is x^0,

$$\frac{\partial J^\mu}{\partial x^\mu} = 0$$

We have energy density and momentum in the four-vector. We have to just put in the four momentum that was built in previous chapters into it.

$$P^\mu = mU^\mu = \frac{E}{c}, \gamma m v(x, y, z)$$

To convert it into a tensor equation, we have to put indices.

First, we have to build energy density and its flow in all three directions.

T^{00} = energy density. First 0 represents energy, second 0 represents density.

T^{01}, T^{02}, T^{03} represent energy flow in three directions.

Next, we form momentum density and flow.

$T^{10}, T^{11}, T^{12}, T^{13}$

This is the momentum density and flow in the x direction.

Similarly, y and z directions indices can be made.

$$T^{\mu\nu} = \begin{pmatrix} T^{00} & T^{01} & T^{02} & T^{03} \\ T^{10} & T^{11} & T^{12} & T^{13} \\ T^{20} & T^{21} & T^{22} & T^{23} \\ T^{30} & T^{31} & T^{32} & T^{33} \end{pmatrix}$$ is the energy momentum tensor.

If we have a dust that is at rest, then the energy momentum tensor will be

$$T^{\mu\nu} = \begin{pmatrix} \rho & 0 & 0 & 0 \\ 0 & 0 & 0 & 0 \\ 0 & 0 & 0 & 0 \\ 0 & 0 & 0 & 0 \end{pmatrix}$$

If there is a dust and momentum flow of particles then in the rest frame, the energy momentum tensor will be

$$T^{\mu\nu} = \begin{pmatrix} \rho & 0 & 0 & 0 \\ 0 & p_x & 0 & 0 \\ 0 & 0 & p_y & 0 \\ 0 & 0 & 0 & p_z \end{pmatrix}$$

We have assumed an ideal fluid, free of particle interactions etc. We can build more complicated energy-momentum tensors.

The continuity equation will now have the form

$$\frac{\partial T^{\mu\nu}}{\partial x^\mu} = 0$$

For this equation to be true for any coordinate system, we have to do covariant differentiation.

$$\nabla_\mu T^{\mu\nu} = 0$$

This could naively be interpreted as conservation of energy and momentum, but we will see later that it's not entirely true, gravitational waves with energy can exist even if the energy momentum tensor is zero.

On the other side of equation, we expect a curvature tensor. Einstein learned about the work done by Reimann on curvature tensors and adopted his work to form the required tensor. Let's study the Riemannian curvature tensor.

The critical thing that determines curvature is how a vector changes if parallel transported around a loop. We discussed this concept earlier. Now we need to measure it precisely.

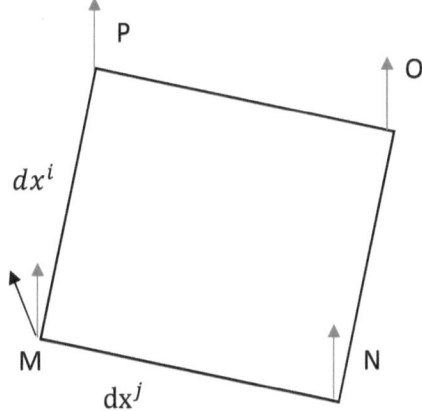

We parallel transport a vector along a loop and calculate change in vector.

Total change in vector going from O to P is integral of change in vector.

$$V_{O-P} = \nabla_j V \, dx^j$$

We need covariant differentiation so that any coordinate system can be used as in the end we want a tensor equation.

Similarly, $V_{N-M} = \nabla_j V \, dx^j$

To complete the loop, we need to calculate difference of differences.

$V_{O-P} - V_{N-M}$ will be along dx^i and differentiation will be ∇_i or $\nabla_i V \, dx^i$

Total change in $V = \nabla_i \nabla_j V dx^i dx^j - \nabla_j \nabla_i V dx^i dx^j$

Doing covariant differentiation twice is not a walk in the park! You could skip this part if not interested in details. We need to do the following calculation

$(\frac{\partial}{\partial x^i} + \Gamma^r_{ki})(\frac{\partial}{\partial x^j} + \Gamma^r_{kj}) - (\frac{\partial}{\partial x^j} + \Gamma^r_{kj})(\frac{\partial}{\partial x^i} + \Gamma^r_{ki})$

Multiplication will lead to pairing of terms

$\frac{\partial}{\partial x^i}\frac{\partial}{\partial x^j} - \frac{\partial}{\partial x^j}\frac{\partial}{\partial x^i} = 0$ term will vanish

$\Gamma^r_{ki}\frac{\partial}{\partial x^j} - \Gamma^r_{kj}\frac{\partial}{\partial x^i} = -[\frac{\partial}{\partial x^i}, \Gamma^r_{kj}] = -\frac{\partial \Gamma^r_{kj}}{\partial x^i}$

Here we used basic product differentiation that $[\frac{\partial}{\partial x}, f]V = \frac{\partial f}{\partial x}V + f\frac{\partial V}{\partial x} - f\frac{\partial V}{\partial x} = \frac{\partial f}{\partial x}V$

Similarly, we get $\frac{\partial \Gamma^r_{ki}}{\partial x^j}$ from the other combination.

Finally, there is term involving only Γ's

In the end, we have $\partial \Gamma$ and $\Gamma \Gamma$ terms left.

Total change in V is represented by Riemannian tensor R given by

$R^r_{kji} = \frac{\partial \Gamma^r_{ki}}{\partial x^j} - \frac{\partial \Gamma^r_{kj}}{\partial x^i} + \Gamma^s_{ki}\Gamma^r_{sj} - \Gamma^s_{kj}\Gamma^r_{si}$

What do these indices mean?

i & j represent a plane formed by coordinates and how it changes as vector moves around in the loop. r & k represent plane formed by the vector itself as it changes directions moving around in a loop. It's hard to visualize the planes in space-time coordinates. So, we rely on mathematics.

Riemannian tensor shows that curvature depends on how connection coefficients change, which in turn depend on how metric tensor changes. This is no surprise as metric tensor tells us the relationship of coordinates from point to point.

In a nutshell, we have a qualitative and quantitative definition of curvature.

A flat curvature is qualitatively defined if a vector does not change direction after parallel transport across a loop.

A flat curvature is quantitatively defined if Riemannian tensor is zero at all points on the space.

We have two sides of a coin. On one hand we have energy momentum tensor which is source of gravitation and on the other we have Riemannian tensor which represents the curvature of space time due to the distortion in the space time caused by the energy momentum tensor.

$R^r_{kji} \sim T^{\mu\nu}$ or

$R^r_{kji} = kT^{\mu\nu}$ where k is some constant.

How do we go about finding out the constant?

We take help from Newton! We already derived Newtonian limit of geodesic equation in the last chapter at slow speeds.

The Gauss law relates gravitational field to the mass density. We derived it in the first chapter. It tells us how a mass spreads its gravitation effect around it.

$\nabla^2 \Phi = 4\pi G \rho$

We can replace mass density by energy momentum tensor as both represent the same thing. The gravitational field ϕ is related to the time-component of the metric tensor as derived in the previous chapter, $\phi = \frac{g^{00}}{2}$.

$$\nabla^2 g^{00} = 8\pi G T^{\mu\nu}$$

$\nabla^2 g^{00}$ means how metric tensor changes. This is what is contained in the Riemannian tensor.

But there is a problem. Riemannian tensor is a rank 4 tensor which means it has 4 indices, but energy momentum tensor has only two indices. This is not allowed as only similar rank tensors are allowed in algebraic calculations.

So, we need to form a rank two tensor. How do we form that?

If you got the hang of indices and tensors, you may have realized that we need to contract indices to get rid of them. But that alone is not enough. There is an additional condition. Remember we had a condition for the energy-momentum tensor that its covariant derivative is zero. This implies energy conservation and continuity equation. Similarly, we impose a condition that the rank two tensor formed from Riemannian tensor should also give zero on covariant differentiation.

$R^l_{\mu\nu m}$ can be contracted by identifying l with m. This means multiplication and summing over all the indices.

$$R^l_{\mu\nu l} = R^1_{\mu\nu 1} + R^2_{\mu\nu 2} + R^3_{\mu\nu 3} + R^4_{\mu\nu 4} = R_{\mu\nu}$$

There we have it, a rank two tensor that would match the energy momentum tensor. It is called **Ricci tensor**.

We impose the condition that $\nabla R_{\mu\nu} = 0$

Unfortunately, if you carry out the covariant differentiation in detail by putting in the formula and all the indices, it's not zero but equal to $\frac{1}{2} g_{\mu\nu} \nabla R$.

What is R?

It's called Ricci scalar and is derived from Ricci tensor by contracting the remaining two indices.

$$R_{\mu\nu}g^{\mu\nu} = R$$

The scalar is like a number, it has no indices and does not change from coordinate to coordinate. In fact, we can do ordinary differentiation on it as its equivalent to covariant differentiation. $\partial R = \nabla R$.

So, we have $\nabla(R_{\mu\nu} - \frac{1}{2}g_{\mu\nu}R) = 0$

This expression can be written in a neat way as

$$G_{\mu\nu} = R_{\mu\nu} - \frac{1}{2}g_{\mu\nu}R$$

$G_{\mu\nu}$ is called the **Einstein tensor**.

Finally, we have the full Einstein field equation

$$G_{\mu\nu} = 8\pi G\, T_{\mu\nu}$$

This is the basic equation of General Relativity. Everything flows from it. We can use upper or lower indices for the equation, it doesn't matter. I have neglected to put speed of light as we were working in units where speed of light is 1 but for completion it goes in the constant term as $\frac{8\pi G}{c^4}$.

Writing an equation does not mean that it's easy to solve. First, there are many equations in it based on the choice of indices. Fortunately, due to symmetries we discussed before, number of independent equations is reduced. Nevertheless, the equations are quite hard to solve. Only in simple cases, they are solvable. Lot of assumptions need to be made to solve them. These days field equations are solved by computers, a field called numerical relativity. Einstein did not solve his own equation. The first non-trivial solution was given by Schwarzschild. We will discuss that solution in detail in the next chapter as that solution leads to the formation of black holes.

A particular problem in solving field equations is that they are nonlinear. It means that both sides affect each other, so relationship between them is complex. It is like saying if we form an equation between politicians and voters, that equation will be nonlinear. The behavior and policies of politicians affect voter choices and their decisions but at the same time, voters' choices influence how politicians react. That's why it is not easy to win elections!

The famous physicist John Wheeler summed it up perfectly when he famously described field equations as matter tells space how to curve and space tells matter how to move.

Let's mathematically analyze the non-linearity.

If we are dealing with vacuum, it means energy momentum tensor is zero.

$T_{\mu\nu} = 0$ or
$R_{\mu\nu} - \frac{1}{2}g_{\mu\nu}R = 0$

We can get rid of indices by contracting with inverse of metric tensor.

$g^{\mu\nu}R_{\mu\nu} - \frac{1}{2}g^{\mu\nu}g_{\mu\nu}R = 0$

$$g^{\mu\nu}g_{\mu\nu} = \begin{pmatrix} 1 & 0 & 0 & 0 \\ 0 & -1 & 0 & 0 \\ 0 & 0 & -1 & 0 \\ 0 & 0 & 0 & -1 \end{pmatrix} \begin{pmatrix} 1 & 0 & 0 & 0 \\ 0 & -1 & 0 & 0 \\ 0 & 0 & -1 & 0 \\ 0 & 0 & 0 & -1 \end{pmatrix} = 1 + 1 + 1 + 1 = 4$$

$R - 2R = 0$ or $R = 0$

$R_{\mu\nu} - 0 = 0$ or

$R_{\mu\nu} = 0$

It is called Ricci flat.

Does that mean there is no curvature of space time?

Nope, zero Ricci tensor does not mean Riemannian tensor is zero. Ricci tensor has less information than Riemannian tensor. It is the Riemannian tensor that determines the curvature and if it is zero then there is no curvature. Zero Ricci tensor still allows for small ripples in space-time, called gravitational waves. They carry energy but do not need any matter sources. This causes non-linearity of Einstein field equations as well as non-conservation of energy.

Let me give you an analogy to make sense of Ricci flatness. Let's create a tensor of stock market

S_i^w

It is made of winners and losers of the stock market. This tensor can be used across various stock markets of the world.

Identifying and contracting indices will result in a scalar S.

$S_w^w = S_1^1 + S_2^2 + S_2^2 = S$

It means we are not interested in information of losers and winners but just want on average, if any money was made in the stock market. This obviously contains less information than the original tensor.

$S = 0$

Means on average, no money was made in the stock market all over the world.

This is like Ricci flat.

Does it mean no one made any money?

Of course not, fat cats always make money! There may be certain stock markets that made money, others lost. This is analogous to saying that Ricci tensor zero does not mean that curvature is zero.

Indices on Riemannian tensor represent movements of coordinate planes and vector planes as vector moves around in a loop. Contracting indices represent identifying plane (two indices) with a single vector (one index).

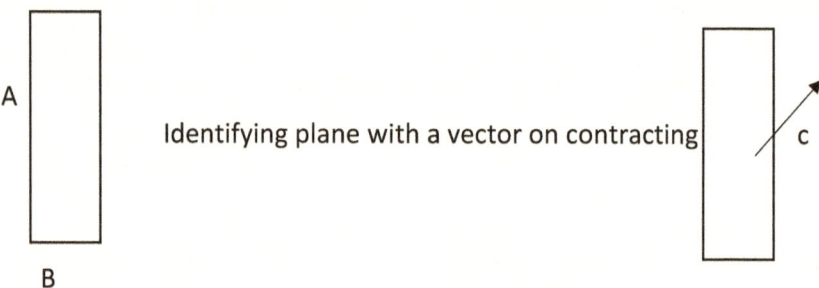
Identifying plane with a vector on contracting

Hope you have some intuitive understanding of the games that are played with indices as it's easy to get lost in mathematical details.

Alternative Approach

Unifying physical theories is the holy grail in physics. It is the ultimate dream of physicists. Unfortunately, physicists have been unsuccessful to unite General Relativity with Quantum Mechanics, another fundamental theory of nature. More on that later but I want to emphasize a mathematical and physical principle that is common to all of these theories and is thus worth learning. This is the Lagrangian mechanics. The equations of Lagrangian formulation lead to correct equations in Classical Mechanics, Quantum Field Theory and General Relativity.

The Lagrangian is defined as Kinetic Energy (KE) -Potential Energy (PE) or T-V. The fundamental principle behind the Lagrangian is the principle of least action. It is one of the most important principles in physics. The principle is simple and elegant. The nature is thrifty. It keeps difference between KE-PE as low as possible.

Action or $S = \int_0^t L dt$

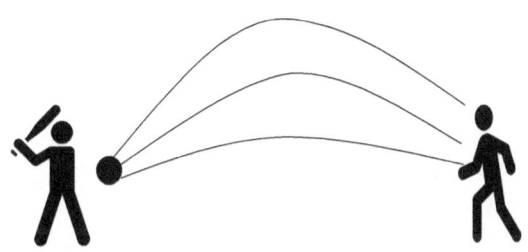

The ball can take several paths to reach the other person at a later time. The path is determined by the principle of least action. The nature chooses the path that keeps the KE-PE difference minimum. So, we integrate over all paths and choose the path where energy difference is minimum. In mathematical terms, we choose the path where action is the least. The technique is part of the calculus of variations.

The equation of motion can be derived from the principle of least action and is called the Euler-Lagrange equation. It determines the trajectory of the ball.

$$\frac{d}{dt}\left(\frac{\partial L}{\partial v}\right) - \frac{\partial L}{\partial x} = 0$$

Let's take the simplest case of the trajectory of the ball falling under gravity
$$L = \frac{1}{2}mv^2 - mgx$$

The potential energy is the gravitational energy given by mgx, where x is the height of the ball. Kinetic energy is $\frac{1}{2}mv^2$ where m is the mass and v is the speed of ball.

$$\frac{\partial L}{\partial x} = \frac{\partial}{\partial x}\left(\frac{1}{2}mv^2 - mgx\right) = -\text{mg=Force}$$

$$\frac{\partial L}{\partial v} = \frac{\partial}{\partial v}\left(\frac{1}{2}mv^2 - mgx\right) = mv$$

mv is the momentum of the ball. This is partial differentiation in action.

$$\frac{d}{dt}\left(\frac{\partial L}{\partial v}\right) = \frac{d(mv)}{dt} = ma = \text{Force}$$

ma=- mg

Obviously a =- g. The ball falls with acceleration caused by the gravity.

Lagrangian formulation was formulated long before anyone knew about Quantum mechanics or General Relativity. It was just a mathematical way of solving Newton's equations. But the Lagrangian has acquired a deeper meaning in modern physics where equations emerge from this formulation in every theory of physics.

How should we go about finding the Lagrangian?

Finding the Lagrangian is like looking for the right talent. There is some guess work involved. This is what professional sports teams do. They look for the right talent that can do the job or play on the field (equation of motion) and score goals. The best talent is someone who can do it with minimum effort and cost, that's your principle of least action!

To find the Lagrangian of General Relativity, it is reasonable to pick something which does not change with different coordinate systems. It could be a rock or earth or sun. In other words, we want a volume element in the space time. The mass of the volume element is useful to put in, after all that is what distinguishes a rock from a star besides the size. This would be the Lagrangian for energy-momentum tensor. For empty space time element, obviously no mass is needed.

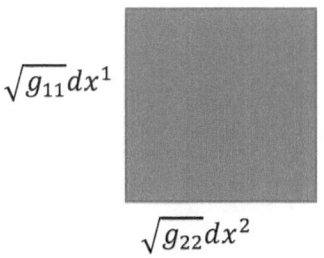

Each length of the volume box is itself a line segment and we know line segment is

$ds^2 = g_{\mu\nu}dx^\mu dx^\nu$ or $ds = \sqrt{g_{11}}dx^1$

Total volume will be multiplying line segments

$dV = \sqrt{g_{11}g_{22}}dx^1 dx^2$

$\begin{pmatrix} 1 & 0 \\ 0 & 1 \end{pmatrix}$ matrix has determinant $1 \times 1 - 0 \times 0 = 1$

This represents area.

This means for a matrix like $\begin{pmatrix} g_{11} & 0 \\ 0 & g_{22} \end{pmatrix}$

$g = |g| = \sqrt{\det g_{\mu\nu}}$ represents the area. det means determinant.

To get the volume we have to include all 4-space time dimensions.

$\sqrt{g}dx^4$ represents volume and is an invariant term. But it does not have any derivatives of g. We want Einstein field equations which are differential equations, in the end. So, we multiply by Ricci scalar R.

Action = $\int \sqrt{g} dx^4 R$

There is no mass density in it, so we have to add a Lagrangian for energy momentum tensor to it.

$L_{mass} = -\int m \sqrt{g_{\mu\nu} dx^\mu dx^\nu}$

We can modify it by multiplying by $\frac{d\tau}{d\tau}$ or

$L_{mass} = -\int m \sqrt{g_{\mu\nu} \frac{dx^\mu}{d\tau} \frac{dx^\nu}{d\tau}} d\tau$

Where m is the mass of the any chosen body.

Putting it all together, we have the total Lagrangian, called the **Einstein Hilbert Action**

$S = \frac{1}{k} \int \sqrt{g} dx^4 R + L_{mass}$

Where k is constant term given by $-\frac{1}{16\pi G}$.

We have to vary this with respect to $\delta g_{\mu\nu}$ so that the change is minimal or 0.

$\frac{\delta S}{\delta g_{\mu\nu}} = 0$

A side note, differentiation with respect to a function like metric tensor is called a functional derivative denoted by δ. This is different than usual derivative denoted by ∂x which is a single variable.

So, we have

$\frac{\delta S}{\delta g_{\mu\nu}} = \frac{1}{k} \frac{1}{\delta g_{\mu\nu}} \int \sqrt{g} dx^4 R + \frac{1}{\delta g_{\mu\nu}} L_{mass} = 0$

The next step is to carry differentiation with respect to determinant g, Ricci tensor and mass term. The detailed steps are complicated with advanced mathematical techniques involved. I will skip the details. Bottom line is, we get back the Einstein field equations, $G_{\mu\nu} = 8\pi G\, T_{\mu\nu}$ in the end.

Einstein's ideas spread before he could come up with the field equations. There was a race to the finish. David Hilbert, a mathematician came up with the action formula about the same time as Einstein published the field equations in 1915, but he acknowledged that the intuition behind this work belongs to Einstein and he gave Einstein full credit for it.

$\dfrac{\delta S}{\delta g_{\mu\nu}} = 0$ is variational calculus.

Think of like it, $\dfrac{\delta volume}{\delta knob} = 0$

If you want to adjust the volume to exact comfort level, you are doing variational calculus. Moving knob widely is going to cause loud and low noise. You want to fine tune the knob so that tiny movements of the knob get closer to the comfort noise level you desire. This is the crux of variational calculus used in the principle of least action.

Gravitational Waves

Let's dig deeper into gravitational waves as they are an important property of General Relativity. Before studying gravitational waves, I want to do a quick review of what is a wave equation and the solution of the classical wave equation. Only then you will be able to recognize formula for gravitational waves.

The classical formula for wave equation for any function f is

$$\frac{\partial^2 f}{\partial t^2} = \frac{\partial^2 f}{\partial x^2}$$ or

$$\frac{\partial^2 f}{\partial t^2} - \frac{\partial^2 f}{\partial x} = 0$$

This is a wave moving in the x direction. Other directions can be added as well. The wave equation can be derived from a vibrating string.

A short hand notation can be used for the wave equation

$$\Box = \frac{1}{c^2}\frac{\partial^2 f}{\partial t^2} - \frac{\partial^2 x}{\partial x^2} - \frac{\partial^2 y}{\partial y^2} - \frac{\partial^2 z}{\partial z^2}$$

$$\Box f = 0$$

Each wave has a frequency. Frequency means how fast a wave is wiggling. The frequency is related to the wave number k.

$$k = 2\pi f \text{ or } \frac{2\pi}{\lambda}$$

Angular frequency is how fast the cycle is being completed.
Angular frequency $\omega = \frac{2\pi}{T}$

The wave number represents how many waves can be packed over 2π. A wave completes its full circle over 2π.

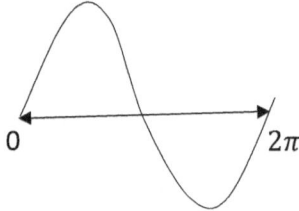

The wave number is 1. But if we can fit three waves into it then wave number will be 3.

The height of the wave is called the amplitude, A.

What is the solution of the wave equation?

It is of the form

$y = A\sin(kx - wt)$

Let me spend some time to tell you how we get the wave equation.

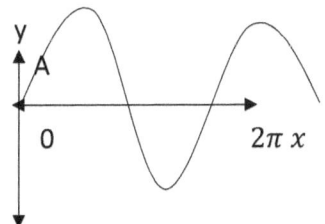

As you can see, the amplitude A on the y axis repeats itself after $\frac{2\pi}{\lambda}$ distance on the x axis. This is like θ. The function can be sin or cosine, doesn't matter as the difference is only in the starting point.

$y = A\frac{2\pi}{\lambda}x = A\sin\theta$ or $A\sin kx$

What if the wave is moving to the right?

Let's take a real-world example. You measure the length of a playing field to be 100m. A person who is going at 10 m/s speed to one end of the field will measure the field to be 100-10=90 m at one second mark. For him or her, the length of the field is $x - vt$.

Similarly, for a wave travelling to the right, the solution changes by
$$y = A\sin(kx - wt)$$

The sine wave can be written in the exponential form using Euler's formula
$$y = Ae^{i(kx-\omega t)}$$

Hopefully you have some idea now how to analyze gravitational waves.

Ok, we will work with weak gravity and in a vacuum. This will simplify things.

Flat space can have small ripples in space-time. This will cause the metric to deviate from the flat space Minkowski metric by a little bit.

$$g_{\mu\nu} = \eta_{\mu\nu} + h_{\mu\nu}$$

$h_{\mu\nu}$ is the small deviation caused by tiny ripples of space-time. We will ignore quadratic terms like $h_{\mu\nu}h_{\mu\nu}$. This is standard way of ignoring very small quantities e.g. if $h_{\mu\nu} = 0.001$ then $h_{\mu\nu}h_{\mu\nu} = 0.00001$ is too small to worry about.

We earlier got the condition from the Einstein field equations that for vacuum where energy momentum tensor is zero, we get

$$R_{\mu\nu} = 0$$

Ricci tensor is made of connection coefficients which intern are made of metric tensor.

$$R_{\mu\nu} = \Gamma - \Gamma - \Gamma\Gamma - \Gamma\Gamma$$

I won't bother with indices here.

$$\Gamma = \Gamma^x_{yz} = g^{xk}\frac{1}{2}\left(\frac{\partial g_{kz}}{\partial x^y} + \frac{\partial g_{yk}}{\partial x^z} - \frac{\partial g_{yz}}{\partial x^k}\right)$$

If we replace metric tensor value with $\eta_{\mu\nu} + h_{\mu\nu}$ then $\Gamma\Gamma$ terms will be quadratic in $h_{\mu\nu}$ that we can ignore, the rest will simplify to

$R_{\mu\nu} \sim \Box h_{\mu\nu} = 0$ besides some other terms

$\Box h_{\mu\nu} = 0$ is like a wave equation if you recall from the earlier discussion.

There is a problem, this equation does not represent a unique solution. If you do a coordinate transformation and get a new $h'_{\mu\nu}$, that will also satisfy a wave equation. So, you have lots of wave equations. That's not good, we want a unique solution if gravitational waves are real. Some of the solutions are trivial, meaning related to property of coordinates, not to physical reality. We want to get rid of them.

This is not a unique problem. It is seen in gauge symmetry and requires gauge fixing. It is a problem of plenty. Loosely speaking if gauge is a measuring device, it should not affect the outcome. You can choose whatever instrument you want but we have to select the most appropriate one based on the conditions. This is a crude analogy as the problem is mathematical. This problem is not new, classically seen in Maxwell equations that describe light.

Let's see how it works in case of electro-magnetic waves as this will give you the idea how it's mathematically done.

How to generate a magnetic field B?

By taking curl of the vector magnetic potential A.

$B = \nabla \times A$

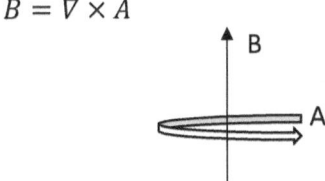

How to generate an electric field?

$$E = -\nabla \phi - \frac{\partial A}{\partial t}$$

Here ϕ is the scalar field. So, potentials lead to fields

In special relativity, we combine space and time
$$A^\mu = \begin{pmatrix} \phi \\ A \end{pmatrix} \text{ or } A_\mu = \begin{pmatrix} \phi \\ -A \end{pmatrix}$$

How would A change if we want to go from one direction to another?
$\partial_\mu A_v - \partial_v A_\mu$ where μ and v are different directions.

This is called the Electro-magnetic tensor, $F_{\mu v}$

$F_{\mu v} = \partial_\mu A_v - \partial_v A_\mu$ where μ and v run from 0(time) to 3(space).

$\phi = \phi$ + any constant is a gauge.

Adding constant does not change anything. The differentiation of a constant is zero, so it does not contribute anything to the equation.

$A = A + \nabla f$ is a gauge as well.

Adding a gradient, ∇f does not contribute to the magnetic field as curl of a gradient is always zero.

$B = \nabla \times (A + \nabla f)$ but $\nabla \times \nabla f$ is zero.

Gradient means going uphill or downhill. Curl means moving in a circular manner and reaching the starting point like a circle. You cannot go downhill and reach the place where you started!

$\nabla . A = 0$ is called a Coulomb gauge. It means no divergence or spreading out.
A field looks like

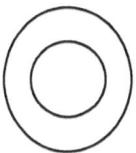

Adding time dimension to the Coulomb leads to the Lorentz gauge.

$$\nabla \cdot A + \frac{\partial \psi}{\partial t} = 0$$

Or in relativistic notation $\partial_\mu A^\mu = 0$

These gauges are used in equations to make calculations simpler.

To fix the gauge in gravitational waves, we invent a new quantity

$$\bar{h}_{\mu\nu} = h_{\mu\nu} - \frac{1}{2}\eta_{\mu\nu}h$$

Then impose the condition to fix the gauge

$$\partial_\mu \bar{h}^{\mu\nu} = 0$$

And the wave equation becomes

$$\Box \bar{h}_{\mu\nu} = 0$$

I know it's a difficult topic. Let me give you another analogy. If a company wants to make a car, they set technical requirements first like how much weight or horse power it produces. The design department has lot of freedom to design the car. They could design whatever they want, a family sedan to a two-door coupe or some weird looking thing. They have to fix the design in the end to meet their mandate. This curtails their freedom and gives a unique product. This is what gauge fixing does by curtailing gauge freedom.

The solution is similar to wave equation solution

$$f = A\sin(kz - wt)$$

Where we replace function and amplitude with the small correction

$$h_{\mu\nu} = h^0{}_{\mu\nu}\sin(kz - wt)$$

If we think of a wave as transverse wave meaning if wave is moving along z axis, then wave is wiggling in x and y planes. It is similar to light waves where a light wave moves along z axis, but electric and magnetic fields are in x and y planes.

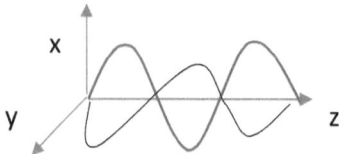

This requires imposition of additional constraints.

It means only x and y components of $h_{\mu\nu}$ are non-zero.

h_{00}, h_{33}, h_{30} etc. will be zero. h_{11}, h_{22} are non-zero.

There is an additional constraint that trace of $h_{11} + h_{22} = 0$.
This means if $h_{11} = 1$ then $h_{22} = -1$. It is called transverse traceless (TT) gauge.

In fact, we can create a matrix

$$h_{\mu\nu} = \begin{pmatrix} 0 & 0 & 0 & 0 \\ 0 & 1 & 0 & 0 \\ 0 & 0 & -1 & 0 \\ 0 & 0 & 0 & 0 \end{pmatrix} f\sin(kz - wt)$$

What effect does it produce?

We know proper distance is related to the metric

$$ds^2 = g_{\mu\nu}dx^\mu dx^\nu$$

If the metric is changing due to gravitational wave passing, the proper distance will change.
e.g. in the x direction, it will increase but since y direction has opposite sign, it will decrease. The sine wave changes sign as it moves with time, so as time passes proper distance along y will increase and decrease along x axis.

This will result in squeezing and stretching of objects as gravitational wave passes through.

Can we detect gravitational waves?

It was thought that the precision needed to observe gravitational waves was so absurd that they will never be detected. It is because the length change between two objects is of the order of 1000th the size of a diameter of proton. But long behold, in 2016 gravitational waves were detected at LIGO and Virgo observatories.

Their detection not only gave a strong vote of confidence to General Relativity, but it is also a triumph of technological advance and scientific collaboration. People involved in the experiment like Kip Thorne, Rainer Weiss and Barry Barish were given very deserving Nobel prizes in physics.

The detection is done through a laser interferometer. A light beam is split into two arms which are several km long. The light is reflected off a mirror and then allowed to interfere. This results in an interference pattern. A gravitational wave will cause distance between mirror and beam splitter to change in two arms. This causes interference pattern to change, which is how gravitational waves are detected.

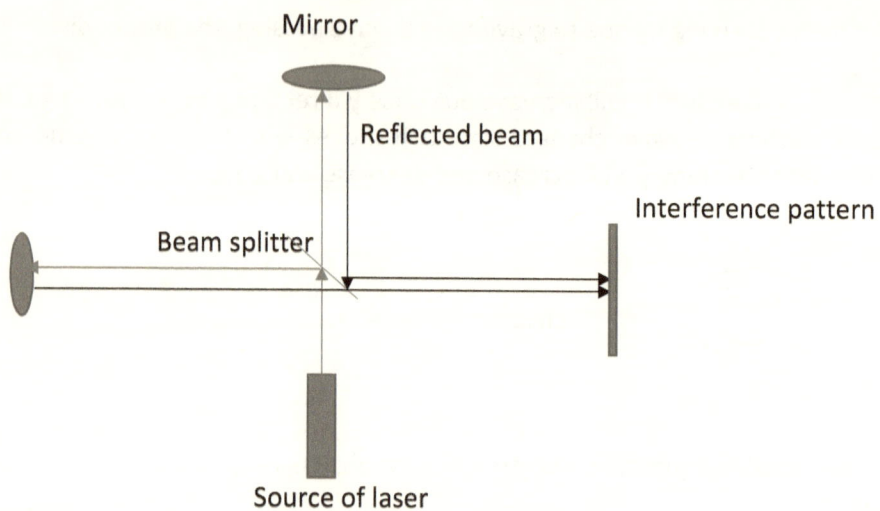

The source of gravitational waves are big astronomical events like merging of rotating black holes that creates huge energy gravitational waves, which can be detected far away on earth. This has opened a new eye for observational astronomy.

Chapter 8

Black Holes

Black holes are a fascinating consequence of the solutions of Einstein's field equations. They represent the extreme curvature of space time. Black holes have captivated human mind ever since they were postulated. They put science fiction to shame. Science fiction has borrowed from General Relativity to imagine the consequences of black holes on human body. We need to focus on how to get to the solution of Einstein's field equations mathematically rather than relying on pop culture clichés about black holes. The first exact non trivial solution to Einstein's field equations was found by German physicist Karl Schwarzschild few months after Einstein published his field equations in 1915.Einstein only had approximate solution which he used to explain the orbit of Mercury which was a major problem that Newtonian gravitation was not able to solve. Interestingly Schwarzschild was posted on the Eastern front during world war I when he found the solution. It's hard to fathom that he had the focus and dedication to solve Einstein's equations when one is not even sure if he is going to be alive at the end of the day! Nevertheless, he sent his solution to Einstein who was amazed that such a simple solution could be found. The concept of black holes was not apparent at that time, in fact it took decades to come to the conclusion that there is a singularity in the solution where space time collapses to a point leading to extreme curvature. The term black hole was popularized in 60's by physicist John Wheeler.

The Schwarzschild solution focuses on the solution outside of a spherically symmetrical body. This makes sense to try as a first solution as most of the astronomical bodies are spherical and we want to see how they curve the space time around them.

Outside vacuum solution

It's obvious to use spherical coordinates since we are dealing with a spherically symmetrical body. There are other simple assumptions that are made like no time dependency of coordinates. We want the body and metric to be static. We do not want influence of other bodies nearby. So, we assume that as we move away from the body, curvature of the space time tends to become flat. Technical term for it is asymptotically flat.

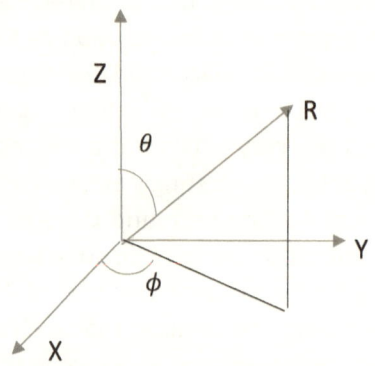

$x = r \sin\theta \cos\phi$
$y = r \sin\theta \sin\phi$
$z = r \cos\theta$

The line segment is

$dl^2 = dr^2 + r^2 d\theta^2 + r^2 \sin^2\theta d\phi^2$

When we include time, then the proper distance becomes

$$ds^2 = g_{\mu\nu}dx^\mu dx^\nu = f_1 c^2 dt^2 - f_2 dr^2 - r^2 d\theta^2 - r^2\sin^2\theta d\phi^2$$

f_1 & f_2 are some functions of radial coordinate. Since we assume solution to be symmetrical all around, dependency on angle θ & ϕ is not needed.

The metric becomes

$$g_{\mu\nu} = \begin{pmatrix} g_{00} & g_{01} & g_{02} & g_{03} \\ g_{10} & g_{11} & g_{12} & g_{13} \\ g_{20} & g_{21} & g_{22} & g_{23} \\ g_{30} & g_{31} & g_{32} & g_{33} \end{pmatrix} = \begin{pmatrix} f_1 c^2 & 0 & 0 & 0 \\ 0 & -f_2 & 0 & 0 \\ 0 & 0 & -r^2 & 0 \\ 0 & 0 & 0 & -r^2\sin^2\theta \end{pmatrix}$$

Our task is to find the functions f_1 & f_2. They can't be hard to find! Nothing is easy in relativity calculations. Let's see what we have to do to find them. We will work at a conceptional level as detailed derivation is cumbersome and will fill half the book.

Start with the field equation.

$$G_{\mu\nu} = 8\pi G \, T_{\mu\nu}$$

If we are dealing with vacuum outside the spherical body, it means energy momentum tensor outside is zero.

$T_{\mu\nu} = 0$ or
$$R_{\mu\nu} - \frac{1}{2}g_{\mu\nu}R = 0$$

We can get rid of indices by contracting with inverse of metric tensor.

$$g^{\mu\nu}R_{\mu\nu} - \frac{1}{2}g^{\mu\nu}g_{\mu\nu}R = 0$$

$$g^{\mu\nu}g_{\mu\nu} = \begin{pmatrix} 1 & 0 & 0 & 0 \\ 0 & -1 & 0 & 0 \\ 0 & 0 & -1 & 0 \\ 0 & 0 & 0 & -1 \end{pmatrix} \begin{pmatrix} 1 & 0 & 0 & 0 \\ 0 & -1 & 0 & 0 \\ 0 & 0 & -1 & 0 \\ 0 & 0 & 0 & -1 \end{pmatrix} = 1+1+1+1 = 4$$

$R - 2R = 0$ or $R = 0$

$R_{\mu\nu} - 0 = 0$ or

$R_{\mu\nu} = 0$

Here is the task

1. Find all the connection coefficients Γ. This requires putting in the relevant metric tensor coefficients of the metric. Recall that connection coefficients are calculated from the formula

$$\Gamma^1_{00} = g^{11} \frac{1}{2} \left(\frac{\partial g_{10}}{\partial x^0} + \frac{\partial g_{01}}{\partial x^0} - \frac{\partial g_{00}}{\partial x^1} \right)$$

We have to calculate all the possible 40 combinations which are a lot! Fortunately, due to symmetry, independent components are greatly reduced, and we are left with only 9 independent coefficients e.g. Γ^1_{00}, Γ^1_{22}, Γ^3_{13} etc. I do not want to list all and their values. If you feel like, try doing calculation yourself by putting in the coefficients from the metric.

2. Find all the components of the Ricci tensor. Recall the Ricci tensor is made by contracting indices of Riemannian tensor.

$$R^l_{\mu\nu l} = R^1_{\mu\nu 1} + R^2_{\mu\nu 2} + R^3_{\mu\nu 3} + R^4_{\mu\nu 4} = R_{\mu\nu}$$

Each component has to be calculated by using in the formula for Riemannian tensor

$$R^r_{kji} = \frac{\partial \Gamma^r_{ki}}{\partial x^j} - \frac{\partial \Gamma^r_{kj}}{\partial x^i} + \Gamma^s_{ki}\Gamma^r_{sj} - \Gamma^s_{kj}\Gamma^r_{si}$$

Then identifying two indices together in the above formula.

$$R^1_{\mu v1} = \frac{\partial \Gamma^1_{\mu 1}}{\partial x^v} - \frac{\partial \Gamma^1_{\mu v}}{\partial x^1} + \Gamma^s_{\mu 1}\Gamma^1_{sv} - \Gamma^s_{\mu v}\Gamma^1_{s1}$$

This single term again has many combinations based on value of μ & v and also involves summing over index s which includes all indices again.

The other terms like $R^2_{\mu v2}$, $R^3_{\mu v3}$ etc. have to be made as well. So, you can see there are tons of terms. Once you gather these boat load of terms, start putting in the value of connection coefficients calculated in the first step.

Fortunately, due to limited number of non-zero connections, there are only four non zero Ricci tensors left. These are $R_{00}, R_{11}, R_{22}, R_{33}$. Their actual values are a bit complicated so let's leave them.

3. Find Ricci scalar by contracting with metric tensor

$$g^{00}R_{00} + g^{11}R_{11} + g^{22}R_{22} + g^{33}R_{33} = R$$

4. Find the Einstein tensor

$$G_{\mu v} = R_{\mu v} - \frac{1}{2}g_{\mu v}R = 0$$

We will again be left with four non zero Einstein tensors $G_{00}, G_{11}, G_{22}, G_{33}$.

4. The four non-zero Einstein tensors will form four equations which can be solved using algebraic and calculus methods.

The end result of this endless mathematical maze will be the value of unknown functions.

$$f_1 = 1 - \frac{k}{r} \text{ and } f_2 = \frac{1}{1-\frac{k}{r}}$$

Where k is some constant and is determined by the initial conditions namely the spherical body. It should depend on the mass, which makes sense as mass bends space time.

The Schwarzschild metric will become

$$ds^2 = (1 - \frac{k}{r})c^2 dt^2 - \frac{1}{1-\frac{k}{r}} dr^2 - r^2 d\theta^2 - r^2 \sin^2\theta d\phi^2$$

To find the exact value of constant k, we need to go to the weak gravity or Newtonian limit. We already derived that in weak gravity; time component of the metric is related to the gravitational potential by

$$g_{00} = 2\phi + 1$$

And furthermore, the potential itself according to Newtonian gravitation is equal to

$$\phi = -\frac{GM}{r}$$

This means $2\phi + 1 = 1 - \frac{k}{r}$
Or $k = 2GM$
There is c^2 term in the denominator, let's just use c=1 units to eliminate it.

So, the Schwarzschild metric becomes

$$ds^2 = \left(1 - \frac{2GM}{r}\right) dt^2 - \frac{1}{1-\frac{2GM}{r}} dr^2 - r^2 d\theta^2 - r^2 \sin^2\theta d\phi^2$$

What do we do with this metric?

Well, we can measure proper distances and times.

Proper distance = $\sqrt{ds}$

Radial coordinate does not measure real distance. It is just a coordinate. It is a line on a piece of paper. It does not measure curved surface. Similarly, we can use different coordinate system and get different answers. But everyone agrees on proper distance and time as they are measured in rest frame with respect to the event.

If we do not care about angles and time, then we can see that proper distance is related to coordinate r by

$$s = \sqrt{ds} = \frac{1}{\sqrt{1-\frac{2GM}{r}}} dr$$

We can see that proper distance is usually bigger than the coordinate distance. That should not be a surprise as space is curved. A straight line like r coordinate is going to fall short.

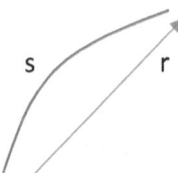

Let's look at proper time. We know that dividing proper distance by c^2 leads to proper time. If we are working with c=1 units then obviously formulae are the same.

$$d\tau^2 = \left(1 - \frac{2GM}{r}\right) dt^2 - \frac{1}{1-\frac{2GM}{r}} dr^2 - r^2 d\theta^2 - r^2 \sin^2\theta d\phi^2$$

If someone is flashing light outside a spherical body like international space station, then proper time is

$$d\tau^2 = \left(1 - \frac{2GM}{r}\right) dt^2$$

Assuming it's a stationary event outside the spherical body then other coordinates are not relevant here.

If r is very large or far away, then
$d\tau = dt$ at $r = \infty$

This means that a very distant observer who is away from the curvature of space time of the spherical body and is thus in a flat space-time will be measuring proper time on his clock that ticks time t.

What about the time measured by the space station that is flashing the light?

$$d\tau^2 = \left(1 - \frac{2GM}{r}\right) dt^2$$

We can see τ is smaller than coordinate time t. But coordinate time t is proper time for a distant observer. This way we can compare proper times for different observers.

$$d\tau_{local}^2 = \left(1 - \frac{2GM}{r}\right) d\tau_{distant}^2$$

Both space station and distant observers are at rest with respect to the events so they can measure proper time, but a distant observer will see the proper time measured by the space craft is slow.

It means a distant observer will see clock ticking slowly on the space station. This is because gravity slows down time. It is called **gravitational time dilation**. So, there is no universal proper time. It is a local concept. Proper time is still invariant, but you have to specify who is measuring it. If you are confused, think of time just like a distance. Since space-time is curved, so local proper time is curved and a straight measuring stick which is only good for a flat space time is going to fall short. It is similar to what we found with proper distance as well.

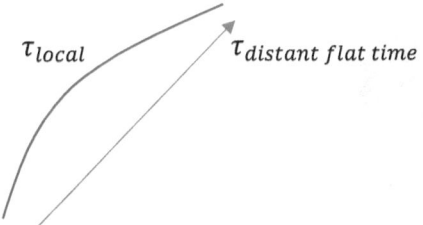

How will a distant observer see light coming from the space station?

Seeing means looking at the color of the light or its frequency. Since frequency is inverse of time period, a distant observer will see higher frequency at the station and lower frequency at his end. In other words, light is red shifted. We already know about this fact from the thought experiments that we started with, but this is an analytic confirmation of it. Think of light coming out of the curved space as trying to climb a mountain, it's going to get tired or lose its frequency/energy.

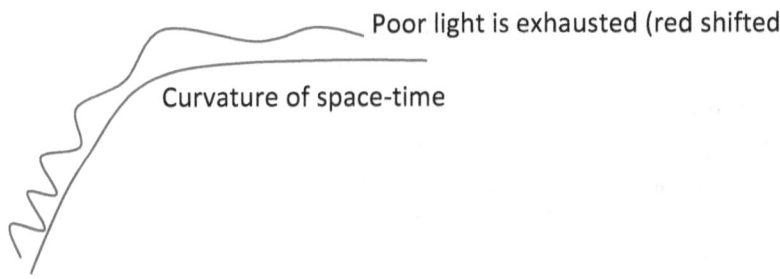

Let's come to the strangest part of the Schwarzschild metric, singularity. It means coordinates become extreme in certain conditions. This happens if coordinates are not properly chosen meaning certain regions are not uniquely represented.
Take example of trying to map a sphere onto a plane surface.

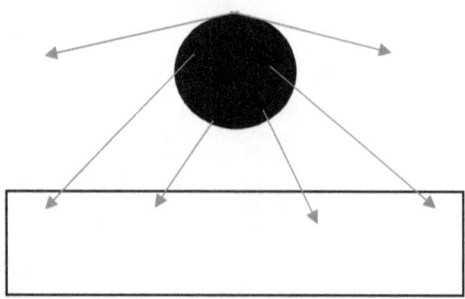

We can see that the poles will not be properly mapped on to the surface as lines stretch till infinity, you could say this is due to inadequacy of the plane or singularity at the poles. So, we have to be careful if singularity is just a coordinate artifact or something physical is happening at the singularity.

Let's see the source of the singularity.

$\left(1 - \frac{2GM}{r}\right)$ term can be written as $\frac{r - 2GM}{r}$

If $r = 2GM$ then the term becomes zero. This is due to coordinate singularity meaning nothing catastrophic is happening physically at $r = 2GM$. We could eliminate this singularity by using different coordinate system. But that does not mean that $r = 2GM$ is trivial. There are important implications. This point is called the **event horizon**. This happens if mass is concentrated in a very small radius. It is not applicable for systems like earth and sun where even horizon will be well below the surface of earth where Schwarzschild solution is not applicable as it only covers vacuum solution. That's why physicists were initially skeptical of the importance of singularities. It was decades later that event horizons and black holes were taken seriously. The event horizon for an earth is of order of few mm.

The event horizon represents point of no return. Once someone enters it, there is no escape. Even light cannot escape. That's why black holes are black. To see this let's look at the proper time.

$$d\tau_{local}^2 = \left(1 - \frac{2GM}{r}\right) d\tau_{distant}^2$$

If $\left(1 - \frac{2GM}{r}\right) = 0$ then

$$d\tau_{local}^2 = 0$$

This means time is frozen at the event horizon for a distant observer. He will see that a person entering event horizon just freezes there, never crosses it. To a person who is entering the event horizon, he sees nothing of the sort, his clock is running normally, and he will cross the horizon in finite time.

The dr term also blows up to infinity.

$$s = \sqrt{ds} = \frac{1}{\sqrt{1 - \frac{2GM}{r}}} dr = \frac{1}{0} dr$$

We know that proper distance is larger than dr as space is curved so straight meter stick, dr will be shorter. But at the event horizon, space is so curved that it is infinitely curved and a straight meter stick will never be able to measure it.

The metric also changes sign inside event horizon.

Outside coordinate time has positive and distance has minus sign, this is because $\frac{2GM}{r}$ is small outside the horizon. But inside the horizon, dt becomes negative and dr becomes positive.

$$d\tau^2 = \left(1 - \frac{2GM}{r}\right) dt^2 - \frac{1}{1 - \frac{2GM}{r}} dr^2$$

Think of this way, your clock always click forwards as future time is inevitable, so we cannot stop time from flowing forwards. But the distance can be reversed, you can move backwards if you want to. But inside the horizon, distance becomes like time coordinate as once you are inside the horizon, the distance to singularity(r=0) becomes the future which cannot be avoided.

At r = 0, there is true singularity which means all terms blow up and there is true collapse of space time.

Do you know the favorite game of black hole physicists?

It's changing coordinates. Each coordinate system has a story to tell about weirdness of black holes. We will focus on Kruskal-Szekeres coordinates as they are commonly used to describe interesting black hole properties.

The main goal of the coordinates is to get rid of coordinate singularity at $r = 2GM$.

The t and r coordinates are replaced by new set of coordinates called T and X.

T is time like and is related to t and r by

$$T = f \sinh \theta$$

X is space like coordinate

$$X = f \cosh \theta$$

Where f and θ are functions of t and r. We don't have to know exact formula and derivation of these coordinates. Let's focus on analyzing what can we learn from them.

What would graph of sinh or coshθ looks like?

They form hyperbolas.

Think of them as accelerated frames.

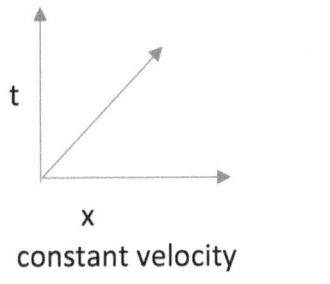

constant velocity

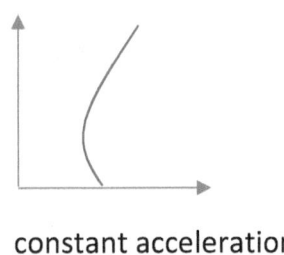

constant acceleration

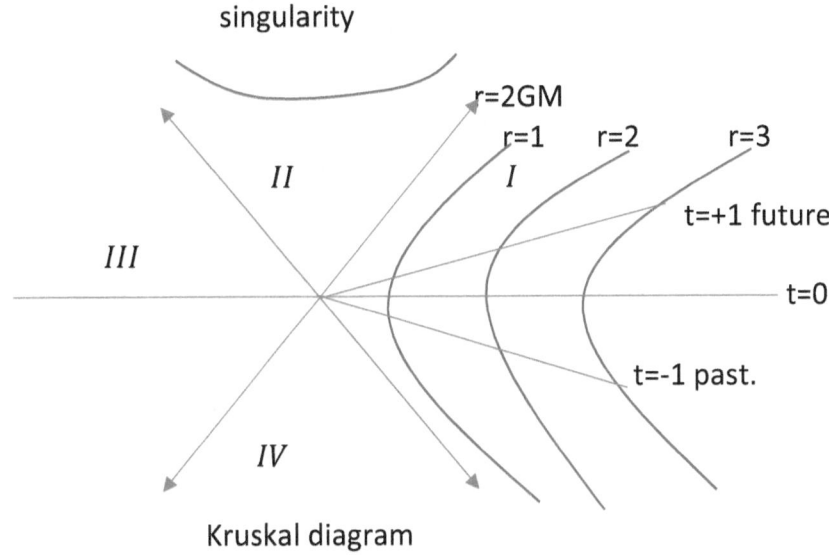

Kruskal diagram

Region I is outside the black hole where $r > 2GM$. It has a past and future.

The intersection of regions is made of event horizon where $r = 2GM$.

Region II is inside the black hole. All world lines or geodesics end in the singularity at r=0.

To see that nothing can exit once inside black hole, we have to solve equations of motion or geodesics which are complicated to solve so just believe that no world lines exit the black hole. Another way of saying is that light cones inside the black hole are all pointing to the singularity.

What about regions III and IV ?

They are not part of the Schwarzschild solution as they represent area within mass distribution of the star that formed the black hole where the Schwarzschild solution is not valid. But nevertheless, if we have to analyze all possible solutions to different values of X and T, we get these regions. This is done by analytic extension of the solution. It is fancy word of saying that we guess the solution where it is not supposed to work.

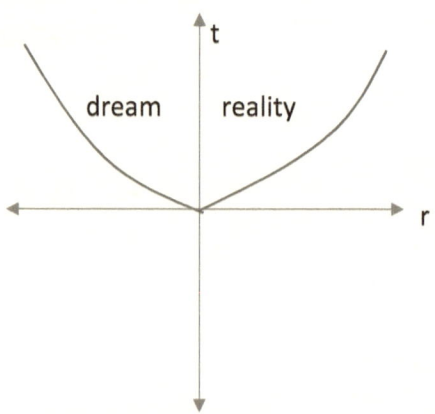

Let's say a function is only defined for positive r. To extend it to negative r, we just assume that the function will behave close to the original function. Think of the dreams, we can extend how we behave in reality to our dreams, this way we can analyze dreams and reality.

Region III represents a past singularity which gave away all its stuff to the present black hole. It has a peculiar characteristic that it can only give away stuff but not get any stuff as opposed to the present singularity which only accepts stuff, does not give away anything (at least classically). It is also called white hole as a result.

Region *IV* represents a space time similar to outside the black hole that flattens out as you move away from the black hole. It is connected to the usual space time at r=GM. This is also called as Einstein-Rosenberg bridge or **wormhole**.

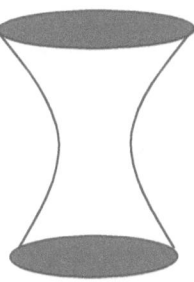

Do *III* and *IV* represent reality?

At present there is no evidence of the existence of wormholes or white holes. Many physicists believe these solutions are just mathematical artifacts, not real world physical solutions. Only time will tell, we can wish that dreams become reality!

There is another way to describe and study black holes and space time. It is through Penrose-Carter diargram. It squeezes all space and time into a finite diagram.

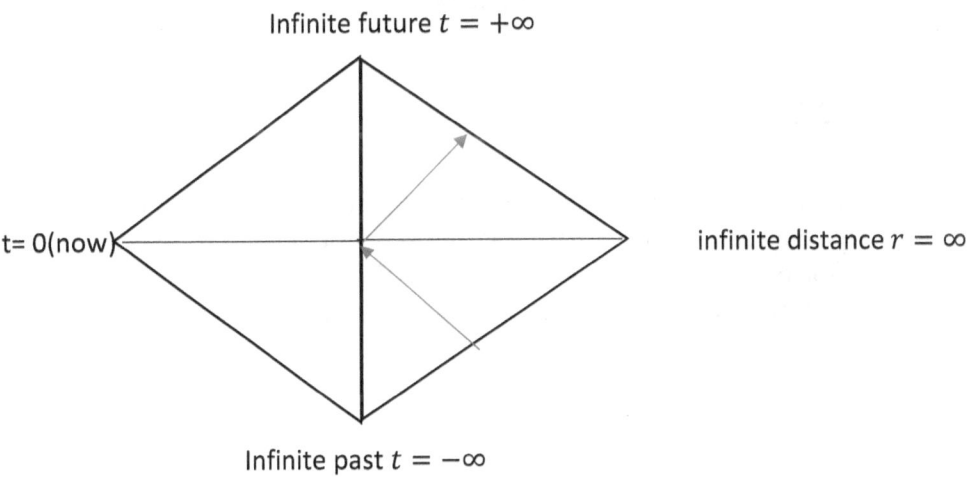

The light rays(in blue) move at 45 degree angles.This means we can draw light cones with them and analyze casuality of events.

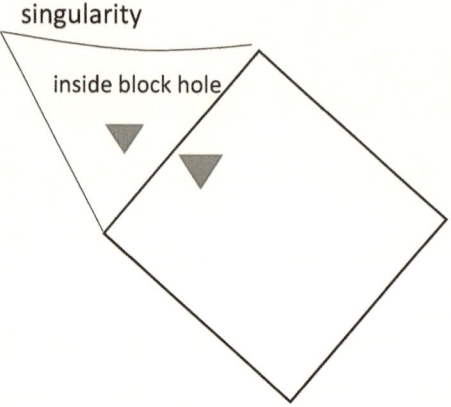

You can see once passed horizon, one has to exceed the speed of light to get out of the black hole.If you are rusty on the details of light cones,go back to previous chapters.

Equations of Motion

To analyze how things move near or inside black holes,we need to solve geodesic equation

$$\frac{d^2x^i}{dl^2} + \Gamma^i_{jk}\frac{dx^j}{dl}\frac{dx^k}{dl} = 0$$

We need to calculate all connection coefficients first, then putting in the value of all four space time coordinates, we will get four geodesic equations. These are quite difficult to solve. There is an easier way to look for equations of motion, that is through symmetries and conserved quantities. Let's first do a brief overview of this concept.

The conservation principles are like guide posts that help particle trajectories to reach their destination. They bring predictability and stability to physics. We are better off with the conservation laws than without them.

What do we mean by conservation?

If a quantity does not change with respect to a variable, we call it conserved.

If a ball on the table is not moving, it is not changing with respect to time.

$$\frac{d(ball)}{dt} = 0$$

The position of the ball(x) is conserved.
If the ball is moving then it will be at a different place at a later time, then
$$\frac{dx}{dt} \neq 0$$

The position is not conserved.

This is known as **Noether's** theorem. It basically says symmetries leads to conserved quantities.

In the language of Lagrangian, if

$$\frac{\partial L}{\partial q} = 0$$

This means if we change q, it does not affect the Lagrangian L. It also means $\frac{dq}{dt}$ does not affect L.

$\frac{\partial L}{\frac{\partial q}{\partial t}} = p$ is conserved.

The momentum is conserved.
In simpler terms, space translation leads to momentum conservation.

Space translation

Whether the table is on one side of the room or another, it does not affect movement of a ball on the table.

If $\frac{\partial L}{\partial t} = 0$

Then the total energy is conserved.
Taking the example of a ball on the table again. The ball has no kinetic energy, but it has potential energy equal to mgh. The potential energy does not depend on time.
If we measure the height or potential energy of the ball- hourly, daily or weekly, it does not change.
$q(t, hourly) \rightarrow q(t, hourly + daily)$

We can say that time translation leads to energy conservation.

Finally, if $\frac{\partial L}{\partial \theta} = 0$

It is called rotational invariance. The ball on the table will obviously fall on tilting of the table as gravity acts on it. But let's take the example of a fidget spinner and assume that it keeps rotating the same way irrespective of orientation.

$\frac{\partial L}{\frac{\partial \theta}{\partial t}}$ = angular momentum is conserved.

The rotational kinetic energy is equal to $\frac{1}{2}I\omega^2$.

I is the moment of inertia. ω is the angular velocity and is equal to $\frac{\partial \theta}{\partial t}$.

$\frac{\partial L}{\frac{\partial \theta}{\partial t}} = \frac{\partial L}{\partial \omega} = \frac{\partial}{\partial w}(\frac{1}{2}I\omega^2) = I\omega$, which is the definition of angular momentum.

The rotational invariance leads to the conservation of angular momentum.

We saw that time independence leads to energy conservation. Well, Schwarzschild metric is time independent. There is a mathematical way of expressing it. If there is conservation of a quantity in a direction, we define a Killing vector in that direction. It is conserved along a geodesic. There is nothing to fear about the Killing vector, it is named after mathematician Wilhelm Killing.

For time, Killing vector = (1,0,0,0) based on four space time directions.

$K \cdot u =$ is conserved, where u is the four velocity.

For time, we put in the time coefficient of the Schwarzschild metric to get the conserved quantity which is energy.

$$\left(1 - \frac{2GM}{r}\right)\frac{dt}{d\tau} = \frac{E}{m}$$

The energy conserved is energy per unit mass.

Similarly, there is symmetry of rotation in Schwarzschild metric

For ϕ, Killing vector = (0,0,0,1) based on four space time directions.

The conserved quantity is the angular momentum, l per unit mass.

$$r^2 \sin^2\theta \frac{d\phi}{d\tau} = \frac{l}{m}$$

There is another equation that we can use which was derived in previous chapters when we introduced relativistic notation.

$$g_{\mu\nu} U^\mu U^\nu = 1$$

Actually its equal to c^2 but we are working in c=1 units.
Putting in the values of conserved quantities into the above equation, we get freely falling equation, assuming we are in the orbital plane so that $\theta = 90$, so $\frac{d\theta}{d\phi} = 0$.

$$\left(\frac{dr}{d\tau}\right)^2 + \frac{l^2}{(mr)^2}\left(1 - \frac{2GM}{r}\right) - \frac{2GM}{r} = \left(\frac{E}{m}\right)^2 - 1$$

$\frac{dr}{d\tau}$ is like a kinetic term, rest of the left side equation is like a potential and the right side could be considered the orbital energy. This helps us analyse orbital motion of a freely falling body around Schwarzchild metric. It shows that orbit of bodies rotates with time in a plane forming a rosette pattern.

We can also calculate the time taken to reach singularity from the event horizon. We assume no angular momentum, so radially falling body, then the above equation simplifies to

$$\left(\frac{dr}{d\tau}\right)^2 = \left(\frac{E}{m}\right)^2 - 1 + \frac{2GM}{r}$$

We have to solve an integral and if we take $r \to 0$ then we get a finite answer. The derailed solution is complicated, but the answer is remarkably simple.

$$\Delta \tau = \frac{4}{3} \frac{GM}{c^3}$$

A body passes through the event horizon uneventfully to reach singularity. Un eventfully does not mean that humans will not feel anything near event horizon. It is not a comfortable place to be! The tidal forces are so strong that there is squeezing and stretching going on as discussed in gravitational waves which leads to spaghettification. It is like squeezing toothpaste out of the container. This is because gravitational field differs greatly from head to toe. In relativity terms, geodesics diverge causing tidal forces.

Photon sphere

Light is bent by the curvature of the black hole. There is a radius at which it orbits the black hole, it is called the photon sphere.

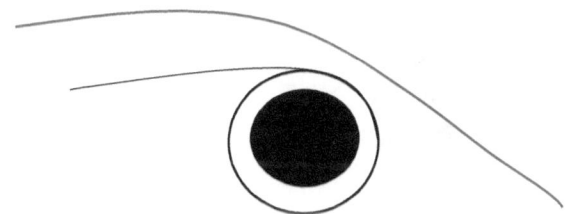

To derive the equation, let's look at the metric

$$ds^2 = \left(1 - \frac{2GM}{r}\right) dt^2 - \frac{1}{1-\frac{2GM}{r}} dr^2 - r^2 d\theta^2 - r^2 \sin^2\theta d\phi^2$$

For photon sphere, $ds^2 = 0$ and its orbiting at fixed radius, so $dr, d\theta = 0$

$$0 = \left(1 - \frac{2GM}{r}\right) dt^2 - r^2 \sin^2\theta d\phi^2$$

$$\frac{rd\phi}{dt} = c\sqrt{1 - \frac{r_s}{r}}$$

Where $r_s = = \frac{2GM}{c^2}$ is the Schwarzschild radius.

$\frac{rd\phi}{dt}$ is like angular velocity which in Newtonian physics is $v = r\omega$.

You can either use Newtonian physics or more appropriately solve the orbital geodesic equation to get the value of r, skipping some steps, we get

$r = \frac{3r_s}{2} = \frac{3GM}{c^2}$. This is the radius of photon sphere.

Rotating Black Holes

It is believed that rotating black holes are much more common than stationary ones. The solution to rotating black holes is called Kerr's solution. Obviously, it is much more complicated but let's focus on its qualitative salient features.

As black hole rotates, it drags the surrounding space time with it. This is called frame dragging. There is a part of the dragged frame where no object can remain stationary or move against the flow of the dragged space time. This is called the ergosphere. It is not the event horizon. So, it is still possible to get out of the ergosphere. But everything rotates along with the direction of the flow of space time. Even light has to travel in the direction of the dragged space time. An object can extract this rotational energy out of the ergosphere and get out of it, the process is called Penrose process. This can lead to depletion of black hole's rotation as it loses angular momentum as a result of it.

There are other solutions which involved charged rotating and non-rotating black holes. They are believed to be much less common in cosmology, so let's leave them.

Astronomical Black Holes

How are black holes formed?

The most common form of black hole is called stellar black hole. It is formed when stars collapse. Most stars after they have run out of fuel, do not form black holes. They burn off their fuel and remaining material forms a dim looking white dwarf. This is because the remaining material is not dense and due to Pauli exclusion principle, electrons cannot occupy same quantum state. So, they exert resistance to gravitational collapse through degeneracy pressure that keeps them away from forming black holes. If remaining material is denser, they can form neutron stars but again neutron degeneracy pressure prevents complete gravitational collapse. A massive star explosion like a supernova is a good candidate to produce massive remaining material that can form a black hole. There is a limit of mass above which gravitational collapse becomes inevitable. Indian physicist Chandrasekhar was first to recognize such limit. He postulated that if the remaining mass is above 1.4 times mass of the sun, gravitational collapse will occur. This is called Chandrasekhar limit. This limit was further refined by Tolman, Oppenheimer and Volkoff(TOV limit) to give a new limit of 2 to 3 solar masses.

The black holes are defined based on their mass with respect to the sun. The stellar size black holes have size comparable to the mass of the sun. Supermassive black holes may have mass that is million or billion times that of the sun. Mini or even microscopic black holes have been postulated as well. Stellar size black holes are surprisingly common. Supermassive black holes are typically found at the center of galaxies. It is postulated that they played an important role in the structure of the universe. It is not clear how they were formed in the first place.

How are black holes detected?

The obvious problem with black hole detection is that they are literally invisible. Even light cannot escape the event horizon so it's tricky to detect them. The only way to know about their presence is to detect the effect on their surroundings. Those effects are special and have characteristic signatures of a black hole.

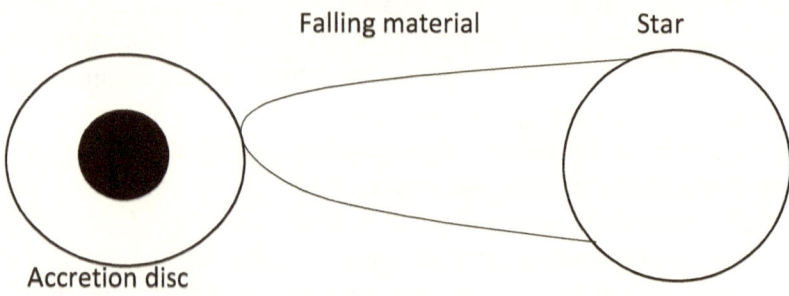

The structure of black hole has been studied in great detail. I will give a simplistic view here. If you want more details, any astronomy and cosmology book will suffice. The details are descriptive so there's no point of me dwelling over those details.

The star in the vicinity of the black hole is sucked into the black hole. The falling material rotates around the black hole and if black hole is rotating, it gains lot of energy and gets very bright. This forms an accretion disc. This can be detected through telescopes. Supper massive black holes eject huge amount of energized radiation in the form of jets. It is thought to be related to generation of magnetic field from rotating black holes. These are called quasars. They are one of the brightest objects in the universe.

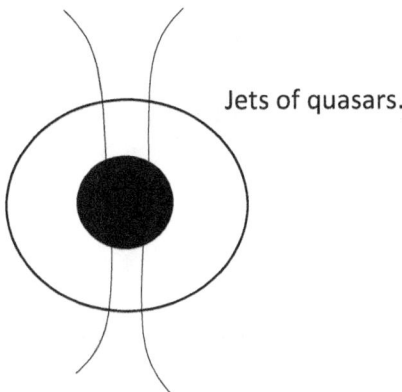
Jets of quasars.

There was plenty of indirect evidence detected by cosmologists over the decades but finally in 2019, direct evidence of black hole observation was made. A direct photograph of black hole with accretion disc was detected by an array of telescopes all over the world (event horizon telescope) from a super massive black hole millions of light years away. So, black holes have become a central theme in theoretical physics and astronomy.

Black Hole Paradoxes

Black holes not only suck matter and heat it up around them forming accretion disc, they also suck physicists into heated debates around them as well. Black hole is a place where clash of titans of theoretical physics takes place. The dominant theory of microscopic matter, Quantum Mechanics is in direct conflict with General Relativity. Physicists including Einstein tried to reconcile both theories together but no consensus on a successful theory has developed in the last 100 years.

A startling discovery was made by Stephen Hawking in his provocative article, Black hole explosions? to Nature magazine in 1974. He used Quantum Field Theory to derive at the radiation emitted by black holes due to quantum uncertainty. The radiation is aptly called Hawking radiation.

The Quantum Field Theory (QFT) is a complicated subject and its application to curved geometry is even trickier. But I can give you a basic outline to at least explain the argument behind Hawking radiation. QFT is a field theory which means fields are the center of action. When fields get excited, they produce particles and antiparticles.

Special Theory of Relativity gives us the energy-momentum relation

$$E^2 = p^2 - m^2$$

Dirac was able to factorize it into $(p + m)(p - m)$ with help of matrices.

If we choose one solution, $E = p + m$

It means energy can be positive or negative. This is because momentum can be positive or negative based on right or left moving waves. The positive energy particles are electrons and the negative energy particles are interpreted as antiparticles called positrons in this case.

QFT Lagrangians contain many terms involving theoretical field ϕ. The theoretical field means, it is for the purpose of mathematics only, it cannot be directly observed. Nevertheless, it gives remarkably accurate experimental results. QFT has a problem of plenty which means every possible interaction is included in the Lagrangian. This means there are self-interacting terms meaning vacuum can interact with itself producing pairs of particles and antiparticles. They are called virtual particles as they cannot be experimentally detected due to their existence for very short time. This is based on another principle called Energy-time uncertainty principle.

$$\Delta E \Delta t \geq \frac{\hbar}{2}$$

ΔE represents spread in the value of calculated energy based on repeated experiments on the same quantum state. But the meaning of Δt is less clear. It refers to a significant change in the mean value of the measurements.

Let me give you an analogy of a ball on the table.

If the ball falls down to the ground, the average value of the position of the ball would change significantly. The time that it takes for the ball to fall to the ground would be Δt.
If the average value is changing fast, Δt is small and vice versa. If you want accurate energy measurements or ΔE to be small, Δt has to be large or average value should be changing slowly.

On the other hand, if average value is changing fast, ΔE would be large. This means we are uncertain about the energy of different quantum states. This can lead to interesting consequences. The most notable being an apparent violation of the classical law of energy conservation. This violation can happen over a very short period of time which is not experimentally detectable. If we are not sure about the energy of the ball on the table and on the floor, then the ball on the floor can make a quick trip to the table and back without any intervention. This would violate the energy conservation as ball on the floor has lower potential energy than the ball on the table.

A virtual round trip!

We can say the ball made the virtual trip as long as Δt is very small. But this seems like a ridiculous assertion as we cannot detect it experimentally. But this is what happens in particle physics. The virtual particles are created out of vacuum, violating energy conservation. But since they last for very short period of time, they are not detectable. They do have indirect affects when calculating the probabilities of various particle physics experiments.

We have particle-antiparticle pairs being produced at the event horizon. If one part of the pair gets trapped inside the event horizon, it is doomed. The other part can fly away from the event horizon.

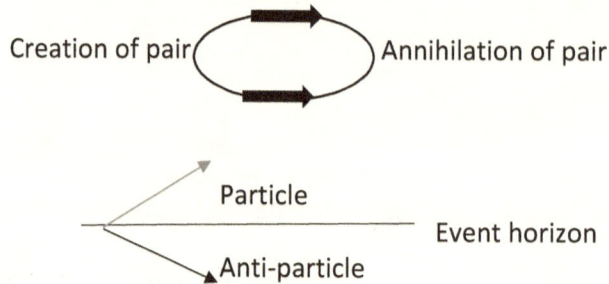

The particle emitted(radiation) carries positive energy as seen by a distant observer. This means the negative energy particle is inside the black hole to conserve energy. The negatively charged particle reduced the mass of the black hole. Remember energy and mass are the same thing, $E = mc^2$. This means as more radiation is emitted, black hole will eventually evaporate. Calculations have shown that time needed for this is huge, many orders more than the age of the universe! The black hole carries a temperature if its emitting radiation. The temperature is inversely proportional to its mass. I will skip the derivation as it's beyond the scope of this book. It is important to remind you that so far there is no evidence of existence of Hawking radiation.

If black hole has temperature, it should have entropy. Let me give you some back ground info about entropy.

Entropy is an extremely important concept in physics. It is at the core of many fundamental principles. It is usually defined in terms of thermodynamics where it represents all possible configurations. It is a measure of disorder of a system. A macrostate in thermodynamics is one that is measured or observed like volume, pressure etc. The microstate on the other hand is all the possible configurations of a macrostate. A material with a particular temperature(macrostate) can be made of various microstates or possible configurations of the particles. If someone is smiling, that's a macrostate. What makes him, or her smile is a microstate.

$$S = k_B \ln W$$

The number of ways(W) in which particles can be arranged is entropy(S). Each state is equally probable.

If we have perfect knowledge of a system, then entropy is zero.
e.g. if a particle can only be in one state then
$S = k_B \ln 1 = 0$

The entropy of an isolated system increases over time. This is an extremely important principle. An isolated system becomes more chaotic over time.
This is called **2nd law of thermodynamics**.

We have all wondered why we can't go back in time. Wouldn't it be wonderful if we can relive our lives or at least good parts of it!

What stops the time from flowing backwards?

There is nothing in Newton's laws of motion, quantum mechanics or Einstein's relativity theory to prevent time from flowing backwards. The answer lies in the 2nd law of thermodynamics. The universe is an isolated system as there is nothing else as far as we know. In an isolated system entropy always increases over time. In an irreversible process, entropy always increases. This means the time in the universe cannot go backwards. Time flowing backwards will decrease entropy, which is not allowed.

Why does entropy always increase in an isolated system like universe?

In other words what is the basis behind the 2nd law of thermodynamics. We have already seen that entropy is the measure of various possibilities. The number of possibilities always increases. If this can happen, then that can also happen, and the game goes on forever. There is nothing to stop the growing of possibilities. This causes entropy to increase and thus fixes the direction of time.

The universe started with a big bang, in a state of low entropy. It started from a point then expanded to form black holes, galaxies, stars, planets and ultimately life.

The universe is getting more complicated with passage of time. The entropy of the universe is increasing, and the arrow of time is one directional. Do not confuse our everyday use of word chaos with entropy. Think of number of possibilities as a measure of entropy. The formation of stars from dust clouds does not decrease entropy as number of possibilities keep growing. Another common mistake is saying that when we clean up our room, don't we decrease the entropy? After all we have arranged things in order. The mistake is we have not considered the work needed. We used our brain, used a vacuum machine to clean the room which taken together increases the entropy.

How can black holes have entropy?

Black holes are singularities, every particle thrown in enters the singularity and loses its distinction as a specific particle. This means singularity is a perfectly ordered place where all states collapse into a single state singularity. The black hole is only characterized by its rotation, charge and mass, this is called no hair theorem.

So, entropy should decrease. Not only entropy, there is loss of information as well. Everything that is thrown in from star to spacecraft, forms a single state of singularity. This goes against the fundamental principle in physics called unitarity. It means distinctions are maintained between states, an electron is different than neutron, nature does not lose this information. If we are talking about classic black holes, we may take solace in the fact that even though information may not be accessible to us once something passes event horizon, at least its locked in there. But Hawking radiation changes that too. If black hole evaporates, the stored information inside it will disappear as well. This is called **information paradox**.

You may say if a book or a painting burns down, information disappears as well. It is not true that information is lost in that case as ash particles and fire have that information coded in them, at least theoretically. The ash of a book will be different than that of a car. That's why certain fumes are considered toxic! But in case of black holes, radiation is not due to incoming matter falling but due to quantum fluctuations on the event horizon. This means, radiation carries no information about the incoming matter. Some physicists have conjectured various scenarios where information may be saved. There is no consensus and I won't go into details of those conjectures. The problem is we do not have a combined theory of Quantum Mechanics and General Relativity, so physicists pick and choose certain principles based on their biases and try to get around the lack of combined theory. It is a patch work, and without experimental evidence, difficult to resolve.

One way to resolve the entropy conundrum is to think that for an outside distant observer, all action of black hole is happening at the horizon. The incoming body never passes the horizon but rather spreads out at the horizon and is frozen in time. This means all this information is spread on the event horizon. The other important fact is that event horizon always increases. This happens if incoming matter keeps falling or even if two black holes merge together. This leads to a conjecture that the key to entropy lies on the event horizon. Since the area of event horizon always increases, this means entropy of the black hole always increases in keeping with the second law of thermodynamics. It is not the volume of a black hole that is important but it's the surface of the event horizon that carries the entropy and the information. This is called the **holographic principle**.

Hawking and Bekenstein came up with the formula for the entropy of the black hole and not surprisingly, it is proportional to the area of the event horizon.

$S_{BH} \sim A$

I hope you are now confident about the black hole physics. If you watch a science fiction movie or listen to a leading physicist talk about black holes, which they invariably do, you can contribute constructively to the discussion.

Tests of General Relativity

Physicists were skeptical of General Relativity at the beginning. After all, it was a radical departure from established Newtonian physics that reigned supreme for centuries. Even when experiments proved General Relativity was correct, Einstein was given Nobel prize not for relativity theory but for photoelectric effect, which was a precursor for the development of Quantum Mechanics. I am not going to give a detailed or exhaustive list of tests confirming General Relativity but rather an outline of important and historical tests.

1. The first test that General Relativity passed was explaining the perihelion of the orbit of Mercury. In fact, it provided some motivation for Einstein to pursue GR as Newtonian physics was unable to explain the experimental findings. When Einstein published his equations in 1915 at the Royal Society, he gave an explanation of the perihelion of Mercury through an approximate solution of his field equations.

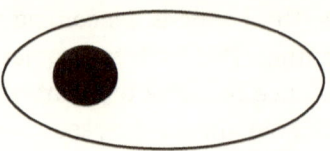

The closest point to the sun is called perihelion. The elliptical orbit of mercury moves about and changes slightly with time as if it is rotating in a plane.

We already saw that the freely falling solution of the Schwarzschild metric

$$\left(\frac{dr}{d\tau}\right)^2 + \frac{l^2}{(mr)^2}\left(1 - \frac{2GM}{r}\right) - \frac{2GM}{r} = \left(\frac{E}{m}\right)^2 - 1$$

Has the potential term $\frac{l^2}{(mr)^2}\left(1 - \frac{2GM}{r}\right) - \frac{2GM}{r}$ which causes the orbit to change.

2. The bending of light by the Sun was a historically important experiment that put Einstein on the world map. The stars that are close to the sun when viewed from earth would appear deflected due to bending of light by the sun. This can be compared to the actual place of those stars from other observations to compare the deflection caused by the light.

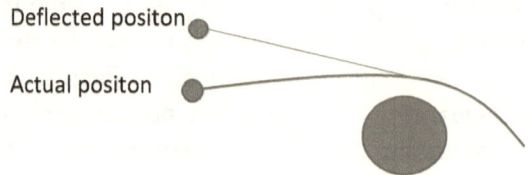

This phenomenon can only be seen clearly during solar eclipse. This experiment was carried out by astronomer Arthur Eddington during the solar eclipse of 1919. There was politics involved in these experiments. Eddington was a pacifist during the first world war. University of Cambridge was the epicenter of pacifists. He was an astronomer at Cambridge and refused to go to the war. He came to know about the new theory of Relativity by a German physicist named Einstein, who himself held pacifist views. This was an opportunity to show that science can transcend nationalistic fervor and show the power of collaboration. This led to the race to prove Einstein's theory. The experiments were correct though accuracy of the data is questionable in hindsight.

3 Gravitational lensing is another example of bending of light and distortion of images caused by curvature of space time.

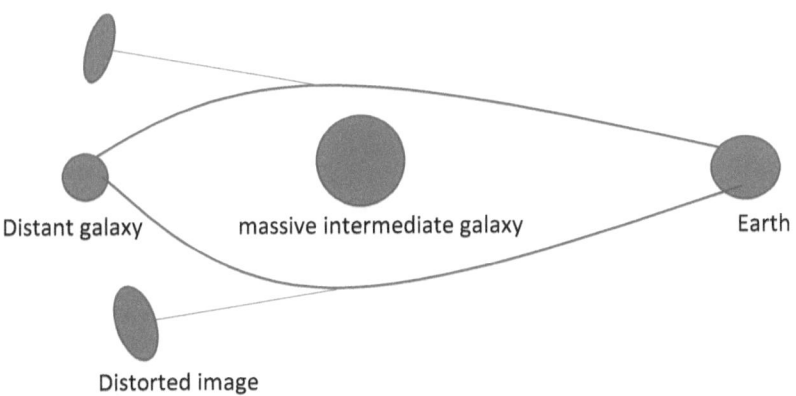

In a simplistic version, a star or a galaxy will appear as two magnified and distorted images or a ring like structure on telescope viewed on earth due to gravitational lensing caused by the intervening mass due to bending of light.

4. The indirect and direct evidence for the existence of black holes has already been discussed. It is strong evidence in favor of General Relativity.

5.Gravitational waves have been detected recently as discussed in previous chapters confirming another pillar of General Relativity.

There is no doubt that General Relativity is on a firm footing with extensive and repeated confirmatory evidence collected over the last 100 years.

Chapter 9

Cosmology

Cosmology deals with study of the universe as a whole. It is not just about counting stars and galaxies. It has transformed from merely an observational science to a branch of physics where prediction of the fate of the universe is made by solving physics equations. The big events that shaped cosmology happened in the early 20th century. The first one was of course the development of General Relativity. The other two events were observational, namely that the universe is same everywhere on a large scale and it is expanding. Universe is same everywhere in every direction is not obvious. We see a complicated structure of the cosmos with galaxies and stars forming various patterns. But if you go to a much bigger scale, of the order of millions of galaxies then universe looks the same. There is no chosen direction, it is same in every direction. The technical name is that universe is **homogenous and isotropic**. The expanding universe suggests that universe is a dynamic place where things are changing all the time. It was the famous American physicist Edwin Hubble who discovered the expanding universe. He observed that the galaxies were red shifted, meaning receding away from us. The farther the galaxy, the faster its moving away from us. This was a shock to the physics community and Einstein. The equations of General Relativity do predict an expanding universe, but Einstein had to introduce cosmological constant to make the universe static. He later called it the biggest blunder of his life. We will discuss the cosmological constant in detail later.

Let's start with a simple model of universe where each block of the universe looks the same. It is convenient to introduce a scale factor to denote the expanding universe.

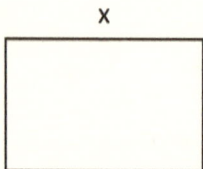

Take a big enough patch of the universe. We could pick any patch in any direction as all patches look the same. If we can analyze it, that applies to all the other patches as well. The distance between two points if we introduce a scale factor (a), is ax. If the universe expands and the distance is doubled, it merely means that the scale factor is doubled. The coordinate x remains the same.

The new distance is $d = 2x$ where $a = 2$.

The velocity will be $\frac{ds}{dt} = v = x\, \partial_t a$. Let's multiply and divide by a.

$$v = ax \frac{\partial a}{a}$$

$$v = Hd$$

H is called the Hubble constant. It is not a constant though as the scale factor does depend on time.

The coordinates do not change, only scale factor changes. You can think of coordinates as points on a balloon, as it expands, the distance between them increases due to stretching of the balloon. The point themselves are the same. This is why these coordinates are called comoving coordinates. The proper distance then becomes

$$ds^2 = dt^2 - a^2 dx^2 - a^2 dy^2 - a^2 dz^2$$

How should we interpret the time dimension?

It is the time that a distant observer would measure if looking at the big patch of the universe expand. It is also called cosmic time.

What about the curvature of space time?

The mathematicians already know what the curvature would look like if the condition is that universe is homogenous and isotropic. There are only three possibilities. The curvature is flat, positively or negatively curved. This can be represented by introducing a curvature parameter, k and giving it values of 0, +1 and -1.

We already studied the peculiarities of curved spaces and the non-Euclidean geometry that comes with it. A plane is a flat surface, parallel lines remain parallel. A sphere is a positively curved surface where parallel lines converge or come close to each other. A saddle is a negatively curved surface where parallel lines diverge or move away.

The mathematicians know how to put the curvature parameter into the metric. The spherical coordinates are used as a convenience and the result is **Robertson-Walker metric**

$$ds^2 = dt^2 - a^2(\frac{dr^2}{1-kr^2} + r^2 d\theta^2 + r^2 \sin^2\theta d\phi^2)$$

You can clearly see if k=0, then the flat metric emerges. However, the flatness is for constant scale factor only. The Riemannian tensor may not be zero if scale factor is not constant. For other values of k, we have to calculate the integral and the proper distance, which will show that the space gets positively or negatively curved.

Einstein field equations

You may wonder that we already got the metric. We do not need to solve Einstein's equations. That's not true, the real dynamic lies in how the scale factor changes with time. This requires solving the field equations.

$$G_{\mu\nu} = 8\pi G\, T_{\mu\nu}$$

Let's introduce the cosmological constant. It is denoted by Λ. It can be added to either side as the solution is still valid.

$$R_{\mu\nu} - \frac{1}{2}g_{\mu\nu}R + \Lambda g_{\mu\nu} = 8\pi G\, T_{\mu\nu}$$

Or you can take it to the other side and make it as a part of the energy momentum tensor. In that case, it represents energy density of the vacuum, also called **dark energy**.

$$R_{\mu\nu} - \frac{1}{2}g_{\mu\nu}R = 8\pi G\, T_{\mu\nu} - \Lambda g_{\mu\nu}$$

Energy momentum tensor of the cosmos

It is not that hard to figure it out. There is mass and there is light. Mass is like dust that forms the energy density (ρ). The light or radiation forms the flow that exerts pressure. The metric will be like

$$\begin{pmatrix} \rho & 0 & 0 & 0 \\ 0 & p & 0 & 0 \\ 0 & 0 & p & 0 \\ 0 & 0 & 0 & p \end{pmatrix}$$

The energy density and pressure contributions come from matter, radiation and cosmological constant. The photons do not have rest mass as there is no rest frame for photons. Its effective mass density can still be calculated based on the energy of photons in a volume.

The density of matter is straight forward. Its related to dimensions of the volume of a box. So, if the box changes in size, which depends on the scale factor then

$$\rho_m \sim \frac{1}{a^3}$$

The density of radiation depends on how many photons are in a volume and how fast are they wiggling. If the box is increased in size, not only density of photons decreases, but their frequency decreases as well, as wave length gets longer to accommodate the bigger volume.

$$\rho_r \sim \frac{1}{a^4}$$

Then we have to know the relationship between energy density and pressure. It is called **equation of state**. It is known from laws of thermodynamics.

$$\rho = wp$$

Where w is a number.

The resting matter does not exert pressure, so w=0. Its value for radiation is $\frac{1}{3}$. This means if the density decreases, pressure is reduced as well.

The value of w in case of dark energy is -1. This means, as density decreases, pressure increases. That means a diluted universe like vacuum, exerts more pressure that pushes things away from each other and is an explanation of accelerating and expanding universe. On the other hand, if density increases, it creates negative pressure, as if it is causing repulsion between matter.

Why does dark energy behave that way?

No one knows. Since we are in the dark about its working, it is appropriate to call it dark energy.

Physicists commonly say that they are not surprised about the existence of dark energy but why its value is so small. The dark energy only comes into play at huge cosmic scales only. The prediction based on QFT is off by order of 10^{-120} which is hugely embarrassing to say the least! This shows how far the two fundamental theories are at reconciling with each other. By comparison difference between Democrats and Republicans on issues is negligible!

Coming back to the energy momentum tensor, the only non-zero components are energy density and three pressure directions. We can also add appropriate metric components from the Robertson-Walker metric to get the following energy momentum components.

$$T_{00} = \rho, T_{11} = \frac{a^2 dr^2}{1-kr^2}, T_{22} = pa^2r^2, T_{33} = pa^2r^2\sin^2\theta$$

Now we need to find the non-zero components of the Einstein tensor and match the components to find equations of motion. It is a labor-intensive process as we saw in calculating the Schwarzschild metric. Here is a recap of the process.

1. Find all the connection coefficients Γ. This requires putting in the relevant metric tensor coefficients of the metric. Recall that connection coefficients are calculated from the formula

$$\Gamma^1_{00} = g^{11} \frac{1}{2} \left(\frac{\partial g_{10}}{\partial x^0} + \frac{\partial g_{01}}{\partial x^0} - \frac{\partial g_{00}}{\partial x^1} \right)$$

We have to calculate all the possible combinations. Fortunately, due to symmetry, independent components are reduced.

2. Find all the components of the Ricci tensor. Recall the Ricci tensor is made by contracting indices of Riemannian tensor.

$$R^l_{\mu\nu l} = R^1_{\mu\nu 1} + R^2_{\mu\nu 2} + R^3_{\mu\nu 3} + R^4_{\mu\nu 4} = R_{\mu\nu}$$

Each component has to be calculated by using in the formula for Riemannian tensor

$$R^r_{kji} = \frac{\partial \Gamma^r_{ki}}{\partial x^j} - \frac{\partial \Gamma^r_{kj}}{\partial x^i} + \Gamma^s_{ki}\Gamma^r_{sj} - \Gamma^s_{kj}\Gamma^r_{si}$$

Then identifying two indices together in the above formula.

$$R^1_{\mu\nu 1} = \frac{\partial \Gamma^1_{\mu 1}}{\partial x^\nu} - \frac{\partial \Gamma^1_{\mu\nu}}{\partial x^1} + \Gamma^s_{\mu 1}\Gamma^1_{s\nu} - \Gamma^s_{\mu\nu}\Gamma^1_{s1}$$

This single term again has many combinations based on value of μ & ν and also involves summing over index s which includes all indices again.

The other terms like $R^2_{\mu\nu 2}$, $R^3_{\mu\nu 3}$ etc. have to be made as well. So, you can see there are tons of terms. Once you gather these boat load of terms, start putting in the value of connection coefficients calculated in the first step.

Fortunately, due to limited number of non-zero connections, there are only four non zero Ricci tensors left. These are $R_{00}, R_{11}, R_{22}, R_{33}$. Their actual values are a bit complicated so let's leave them.

3. Find Ricci scalar by contracting with metric tensor

$$g^{00}R_{00} + g^{11}R_{11} + g^{22}R_{22} + g^{33}R_{33} = R$$

4. Find the Einstein tensor

$$G_{\mu\nu} = R_{\mu\nu} - \frac{1}{2}g_{\mu\nu}R$$

We will again be left with four non zero Einstein tensors $G_{00}, G_{11}, G_{22}, G_{33}$.

4. The four non-zero Einstein tensors can be combined with the corresponding energy momentum tensor components.

The end result is there are two independent equations, commonly called as Friedmann equations.

$$\left(\frac{da}{adt}\right)^2 = \frac{8\pi G}{3}\rho - \frac{kc^2}{a^2}$$

In terms of Hubble constant, it becomes

$$H^2 = \frac{8\pi G}{3}\rho - \frac{kc^2}{a^2}$$

The second equation is sometimes called acceleration equation as it involves second derivative of the scale factor

$$\frac{d^2a}{adt^2} = -\frac{4\pi G}{3}\left(\rho + \frac{3p}{c^2}\right)$$

What to make of the Friedmann equations?

The analogy in Newtonian gravitation is straight forward. Let's imagine a galaxy at the edge of a chunk of the universe. For the purpose of calculation, we can approximate that the rest of the mass of the chunk is located at the center.

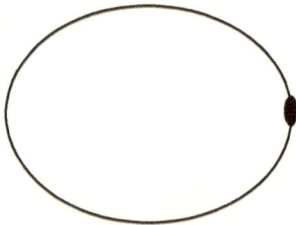

The galaxy possesses kinetic and gravitational energy

$$\frac{1}{2}mv^2 - \frac{GmM}{r} = E$$

$$v^2 - \frac{2GM}{r} = K \text{ , where K is some constant}$$

The velocity in terms of scale factor will be

$$\left(\frac{da}{adt}\right)^2 - \frac{2GM}{ax} = K \text{ , where distance r is expressed in terms of scale factor.}$$

The only thing remaining is to express mass of the region in terms of density and after some clever manipulations, we will get our Friedmann equation back

$$\left(\frac{da}{adt}\right)^2 = \frac{8\pi G}{3}\rho - \frac{kc^2}{a^2}$$

In other words, it's an energy equation with kinetic and potential energies controlled by the constant k.

If k=+1, it means, there is enough escape velocity for the galaxy to fly away.

If k=-1, it means there is no escape velocity and galaxy falls back in.

If k =0, it means galaxy is just at the escape velocity. It can just stay there.

The equation for k=0 becomes $\left(\frac{da}{adt}\right)^2 = \frac{8\pi G}{3}\rho$

And expressed in terms of density and Hubble constant, we get

$$\rho_c = \frac{3H^2}{8\pi G}$$

ρ_c is called the critical density.

Various cosmological models can be made based on different values of density and curvature parameter k.

In simplest case, let's take k =0, meaning a flat universe.

How will scale factor change with time?

The density is made of matter, radiation and cosmological constant components.

In pure radiation universe, we know that $\rho_r \sim \frac{1}{a^4}$

$$H^2 = \frac{8\pi G}{3}\rho$$

Solving this differential equation, leads to

$$H \sim t^{\frac{1}{3}}$$

In pure matter universe, we know that $\rho_m \sim \frac{1}{a^3}$

Solving this differential equation, leads to

$$H \sim t^{\frac{2}{3}}$$

This model is also called Einstein- De Sitter model.

These models expand with time like

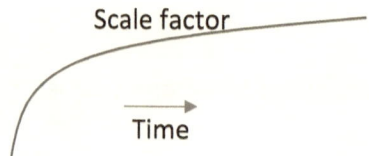

In pure dark energy model, the cosmological constant replaces density parameter.

$$H \sim \sqrt{\Lambda}$$

The solution is an exponential

$$a \sim e^{\sqrt{\Lambda}\, t}$$

The cosmological constant can be absorbed in the Hubble constant parameter and then we get

$a \sim e^{Ht}$

This is an accelerating and expanding universe.

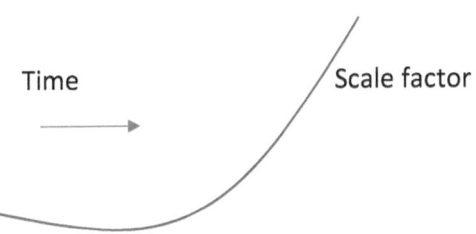

It is useful to create cosmological models based on ratio of density of universe to critical density.

$\Omega = \dfrac{\rho}{\rho_c}$

If we ignore cosmological constant, then its interpretation is straightforward. It helps us determine the curvature parameter k, which tells us about the shape of the universe.

If $\Omega > 1$ then universe is quite dense so eventually gravity will take over. The curvature parameter will be positive, so universe folds like a sphere. This means an expanding universe will eventually re-collapse leading to a big crunch.

If $\Omega = 0$ then density is just matched to the critical escape velocity. This leads to a flat universe.

If $\Omega < 1$ then density is thin, curvature parameter is negative, and universe can expand forever.

The cosmological constant always tends to push things apart so if we add that into the mix, more variety of models can be made, all trying to expand the universe if cosmological constant is large.

Models are fine but which universe do we live in?

The current evidence suggests that curvature parameter for our universe is zero suggesting a flat universe but a positive cosmological constant. This means an expanding and accelerating universe.

Big Bang

The Big Bang is our best possible model for the universe based on the current evidence. Universe started from a singularity. What caused this bang? No one knows. No physical theories exist to answer this question convincingly. The primordial singularity is the start of space and time. We don't expect anything to come out of a singularity as gravity is too strong as per General Relativity, but all theories fail at the start of the universe. The universe then entered the smallest possible configuration. This is called the Planck epoch. The Planck scale is the smallest possible scale known to us so it's natural that universe first enter this scale. It did not last long, lasting only 10^{-43} seconds. The length scale of the universe was of the order 10^{-35} m and temperature was of the order 10^{32} Kelvin. Again, we do not have a theory to explain Planck era. It needs quantum theory of gravity as all the fundamental forces namely strong force, weak force, electromagnetic and gravitational force were of the same order. The forces began to separate starting with gravity. The universe expanded suddenly in a fraction of a second to a size of the order 10^{26}. This is called **cosmic inflation**. This theory was added to the big bang model due to horizon problem. We can only see part of the universe that is in close contact with us, meaning from where we can get light to reach our telescopes. We can see in all directions and there are certain parts of the universe which were never in close contact with one another, but we can see them. The problem arises from the fact that universe was initially very hot, radiation dominated, and it eventually cooled to become matter dominated. The relic radiation from that era is still present and is called **cosmic microwave background radiation**. If you remember Television sets in old days, there used to be static noise if no channel was selected.TV set was actually detecting cosmic microwave radiation. This radiation has been studied in great detail and shows remarkable similarity across all the visible universe. This needed an explanation. How can parts of the universe have the same background radiation when they were never in contact with one another? The cosmic inflation theory explains it as the homogenized universe expanded rapidly. So very distant parts of the universe were in contact with one another before the inflation took place.

Another mystery is the dominance of matter over anti-matter. The anti-matter is like mirror image of matter. It carries complimentary properties of matter e.g. if electron has negative charge then its anti-matter counterpart has positive charge. But in our present universe, there is hardly any anti-matter left. It is hypothesized that both were equally abundant in the early universe and for reasons which are not clear, matter won the race. The technical term for it is baryogenesis. Baryons represent ordinary matter. Once the universe got inflated, it first cooled down then reheated again, where atoms started forming and that led to the end of the radiation era. Once atoms or ordinary matter were formed then structures like stars and galaxies started to develop. The current evidence suggests that our universe is approximately 13 billion years old.

Currently our universe is in the dark energy dominated era where expansion of the universe is accelerating. This evidence came in 1990's when it was found out that distant supernovae are much dimmer than expected. Cosmologists use supernovae as standard candles as they shine the same way. So, a dimmer supernova is farther away. The red shift is measured to calculate the expansion, or the scale factor and the acceleration of the scale factor is then measured to get the expansion rate.

The redshift due to universe expansion is different than due to curvature or doppler shift due to relative speeds. It is like saying that the galaxies are not moving away but the space between them is expanding. It is like dots on a balloon. As the balloon expands, dots are not moving but the space between them has expanded due to stretching of the balloon.

Mathematically red shift parameter is called z.

It is related to the shift of the wavelength by

$$z = \frac{\lambda - \lambda_0}{\lambda_0}$$

Where λ is the wavelength observed and λ_0 is the wavelength that was emitted.

The relationship to the scale factor is important. To find that out, we need to use the Robert Walkerton metric and the null distance of light.

$$ds^2 = dt^2 - a^2 \frac{dr^2}{1-kr^2}$$

$$0 = dt^2 - a^2 \frac{dr^2}{1-kr^2}$$

With integration, we can find the relationship between time and the scale factor when time signal started. Time is nothing but frequency of the radiation. The formula is

$$a(emmision\ time) = \frac{1}{1+z}$$

z is calculated experimentally and then scale factor can be determined which helps us to calculate its acceleration rate.

Physicists were not quick to accept the idea of big bang initially. The fact that universe began from nothing with limited explanation has religious undertones to it. That's why competing theories like steady state model were proposed to explain the expansion. But the cosmic microwave background gave strong evidence in favor of big bang and eventually scientific community accepted it as the de facto explanation of our existence.

There are plenty of details in the big bang model. I do not want to do an exhaustive review as it reads like a story and I don't think I can add any unique perspective to it.

Cosmic Horizon

The Special Theory of Relativity puts a damper on cosmology. Speed of light is the ultimate speed barrier. It is our observational limit. This does not mean that galaxies cannot exceed the speed of light. This is because it is not the galaxies that are moving apart, it is space that is expanding and there is no limit to spatial expansion.

$$v = Hd$$

We can see that there is no limit for the speed of light in this formula.

The farthest things in the universe that we can see now form the **particle horizon**.

The things that we saw or will ever see in the future form the **cosmic event horizon**. This means beyond the event horizon, there are objects that we will never see in the future.

We already know that black hole horizon is such a thing. We cannot see inside the event horizon of the black hole. Similarly, the early radiation dominated universe is invisible as well. It is possible that distant galaxies that we see today will never be seen again in the future as they may have already moved away from us exceeding the speed of light. So, they were inside our particle or event horizon but since then they have left our event horizon.

The Hubble distance is $d = \frac{c}{H}$. The objects inside the Hubble distance are visible to us now or in the future as they are moving away from us less than the speed of light. The objects that exceed the speed of light moving away from us become invisible to us. Now the Hubble constant is not constant, so the Hubble distance is not constant either. In a shrinking universe, Hubble constant will decrease, leading to increase in Hubble distance, so we can see objects that exceed the speed of light as Hubble distance expands. Even in an accelerating universe, as long as Hubble constant decreases, meaning universe is accelerating but at a slower rate, Hubble distance will increase. This way things outside the Hubble distance can come inside.

Let me give you an analogy from movies. The oldest movie that you can see today is like particle horizon. All the movies that you will ever see or can see, form your event horizon. There are movies that have been lost forever, they are outside your event horizon. When you watch a movie, you are looking into the past. The actors are not in the situations shown in the movie at present time. Some of the actors will make more movies and remain part of the event horizon. Other actors in the movie will not work in movies any more, they will exit your event horizon. The new actors into subsequent movies will enter your event horizon in the future and so on. Hopefully this will give you some clarity into cosmic horizons.

References

I am extremely grateful to the resources that provided invaluable information in learning physics. I have listed the important ones for your review and tried not to miss anything inadvertently. If there is any mistake in the list, please let me know.

1. Hartle, James (2003) Gravity: An Introduction to Einstein's General Relativity, Pearson Education Inc., Addison Wesley, San Francisco.

2. Guidry Mike, (2019) Modern General Relativity: Black Holes, Gravitational Waves and Cosmology, 1st Edition, Cambridge University Press, UK.

3. Lambourne, R (2010) Relativity, Gravitation and Cosmology, Cambridge University Press, New York.

4. Serway R, Beichner R, Jewett, J (2000) Physics for Scientists and Engieeners, 5th edition, Harcourt College Publishers, Orlando.

Online Resources

Information is increasingly available online. It is not possible to only use books for learning physics. Online tools are critical in self-study. I don't think it's a surprise for you that every topic gets googled and first checked on Wikipedia. I will list quality resources that I found useful. Since website addresses get changed, I have included the description of the source so that it can be easily searched.

1. Perimeter Institute of Theoretical Physics at Waterloo, Ontario has an excellent outreach program. You can find lots of videos on relativity at perimeterinstitute.ca.

2. Stanford University has YouTube channel. Professor Susskind has video series especially for public, covering topics like quantum mechanics, particle physics, special and general theory of relativity and string theory. These lectures are a treasure. Professor Susskind has a dynamic personality and he is very engaging while teaching physics.

3. MIT has an open course ware website at ocw.mit.edu where there are extensive lectures and videos on topics like relativity, quantum mechanics, particle physics and string theory.

4. Arxiv.org is a repository of publications, run by Cornell University. It has some excellent articles. Lecture notes on General Relativity by Sean Carroll (1997) provide a good but advanced level review of the subject.

5. Khanacademy.org is an excellent online source to learn mathematical topics like calculus, complex functions and series expansions etc.

Index

A
Accretion Disc 136
Action Principle 102

C
Christoffel Symbol 74
Contraction of Indices 64
Contravariant Vector 62
Cosmic Horizon 160
Cosmic Inflation 158
Covariant Differentiation 74
Covariant Vector 62
Critical Density 155
Curvature 47

D
Dark Energy 150
Differentiation 17

E
Einstein Summation 62
Einstein Tensor 95
Equation of State 151
Equivalence Principle 85
Energy-Momentum Relation 43
Energy-Momentum Tensor 90
Entropy 140
Event Horizon 122
Euclidean Geometry 49

F
Friedmann Equation 154

G
Gauss Law 21
Gauge 68
Geodesics 50,79
Gravitational Field 21
Gravitational Waves 103

H
Hawking Radiation 140
Holographic Principle 143
Hubble Constant 148

I
Inertial Frame 24
Information Paradox 142
Integration 12
Intrinsic Curvature 44

K
Kerr's Solution 134
Killing Vector 131
Kruskal Diagram 125

L
Lagrangian 100
Length Contraction 35
Light Cone 27
Lorentz Transformations 38

M
Manifold 50
Matrices 15
Metric 56
Michelson Morley exp. 25
Minkowski Diagram 29

N
Newton's Law of Gravitation 19

P
Parallel Transport 45
Photon Sphere 133
Polar Coordinates 58

Q
Quasars 137

R
Relativistic Energy 40
Relativistic Momentum 40
Ricci tensor 94
Riemannian Tensor 92
Robertson-Walker Metric 149

S
Schwarzschild Metric 118
Simultaneity 26
Singularity 122
Spherical Coordinates 59

T
Tensors 71
Time Dilation 31, 120
Twin Paradox 34
W
Worm Hole 127

www.ingramcontent.com/pod-product-compliance
Lightning Source LLC
Chambersburg PA
CBHW021819170526
45157CB00007B/2641